THE
COSMIC
VERSES

THE COSMIC VERSES

A Rhyming History of the Universe

JAMES MUIRDEN
Illustrated by DAVID ECCLES

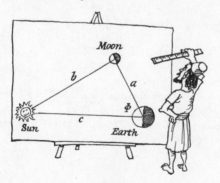

WALKER & COMPANY
NEW YORK

Published by Walker Publishing Company Inc., New York
Distributed to the trade by Holtzbrinck Publishers

All papers used by Walker & Company are natural, recyclable
products made from wood grown in well-managed forests. The
manufacturing processes conform to the environmental regulations
of the country of origin.

Library of Congress Cataloging-in-Publication Data has been applied for.

ISBN-10: 0-8027-1569-9
ISBN-13: 978-0-8027-1569-2

Visit Walker & Company's Web site at www.walkerbooks.com

First published in Great Britain in 2006 by Michael O'Mara Books Ltd.
First U.S. edition 2007

1 3 5 7 9 10 8 6 4 2

Designed and typeset by Martin Bristow
Printed in the United States of America by Quebecor World Fairfield

Contents

General Introduction   11

1 From Stones to Sexagesimals (before 700 BC)   13

2 Lively Minds (c. 700 BC–AD 200)   25

3 Staying Still . . . (200–1200)   43

4 Wearing Spectacles (1200–1500)   59

5 No Fixed Abode (1500–1609)   80

6 The Fall of the Apple (1608–1687)   114

7 Measuring (1671–1838)   143

8 Heaps of Data – Answers Later (1770–1900)   163

9 On to the Beginning (after 1900)   189

Epilogue   223

For Helen

THE
COSMIC
VERSES

General Introduction

It was a most auspicious date
when Mankind learned to *cogitate*!
Birds of the migrant sort, it's said,
have a computer in their head
that monitors their aerial flight
according to the stars (by night),
and use the Sun throughout the day
to guide them safely on their way.
But this amazing fact does not
appear to puzzle them one jot!
You would not, I am sure, expect
a Swift or Swallow to reflect:
'I find this business very odd!
Why don't we all get lost? Some god
must navigate us through the air . . .
we'll thank him – if he gets us there!'
It seems to be a human trait
to analyse and contemplate,
to sort and measure, and to try
to answer questions starting 'Why . . . ?'

1 *From Stones to Sexagesimals*
(before 700 BC)

Created or mutated?

Two million years ago at least,
our ancestors, half-man, half-beast,
were trying out an upright stance.
What started this — Design, or Chance?
The argument's by no means ended;
both views are vigorously defended,
and both may have a point — who knows?
Through many cycles Europe froze,
then melted, dried, and froze once more.
I'll start before the current thaw,
with northern stretches under snow
some twenty thousand years ago.

Out of Africa

Neanderthalers were no fools
(they learned how to use stones as tools),
but found themselves in second place
when challenged by our modern race,
Cro-Magnon man. Our forebears spread
from Africa around the Med,
wherever there was flint (a stone
that they had learned to flake and hone);
and left behind what seem to be
the earliest signs of Artistry.
In caverns near the Pyrenees,
wall-paintings of some expertise

Or *Homo sapiens*
c. 100,000 BC

Or *Homo sapiens sapiens*

c. 10–20,000 BC

depict the animals they chased,
no doubt because they liked the taste.
Although suggesting, to our eyes,
contenders for the Turner Prize,
we rightly contemplate in awe
this quantum leap – this urge to *draw*.
Were these depictions of their prey
thought magical in any way?
Did they appeal to a Creator
to help them make their aiming straighter?
Did furrowed brows contract in vain
(in spite of their upgraded brain)
as they began to wonder why
there is such order in the sky,
concluding that some vaster Force
ordains and regulates its course?

Settling down

Despite their homely decoration,
they didn't stay in one location:
when all the food supplies were gone,
they packed their things and wandered on.
A group of more than 25
would have a struggle to survive
if they just fed on what was growing;
but still, these little bands kept going

till roughly 10,000 BC,
when we find signs of *husbandry* –
tilling the land, and sowing seed.
Another quantum leap indeed,
for once they started settling down,
in no time flat you had a town!
Europe was still too cold and wet
(the ice had hardly left it yet)
for anything to germinate
until a somewhat later date;
and so nomadic ways first ceased
in Egypt and the Middle East.
The Fertile Crescent was an arc
extending westwards from Iraq

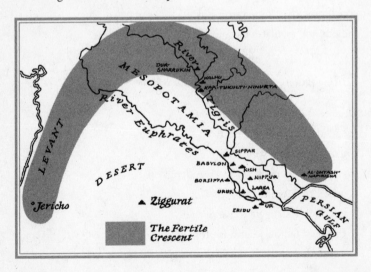

and ending up near Jericho
(the oldest city that we know).
Here, farming methods first took root,
and Egypt quickly followed suit . . .

The Dog Day

The river Nile's annual flood
brought volumes of nutritious mud
stirred up from its tumultuous bed,
which farmers channelled off, and spread
upon their dehydrated fields,
engendering tremendous yields.
The Pharaohs' empire was built
on heaps of this organic silt;
but how could they administrate
without some notion of the date?
When should their New Year's Day occur?
Since this began their calendar,
the whole year's dating would be wrecked
if this one turned out incorrect.
The Dog Star* helped them very nicely,   *Sirius (α Canis Majoris)
allowing them to time precisely
the morning that began the year
(assuming that the sky was clear).
They got a wakeful sharp-eyed priest
to scan the sky, low in the east,

before the Sun came up, until
he had the satisfying thrill
of being able to set eyes on
the Dog Star, low on the horizon.
(By then, of course, it was too late
to have friends round to celebrate.)

This happened (with the naked eye) about the middle of July.

Fact or fancy?

The Pyramids (where they interred
the Pharaohs) earn a passing word.

Seasonal stars

The Earth goes round the Sun each year,
and since it's so extremely bright,
stars are outshone, and disappear
when they are in its line of sight.
They vanish in the sunset glow,
and then a month or two goes by,
until the line of sight moves – so . . .
and they're seen in the dawnlit sky.

Squarely aligned north-south, east-west,
these awesome monuments suggest
some knowledge of astronomy.
But do we find more than we see?
A shaft breaks through the northern face
of Khufu's earthly resting-place . . .   Great Pyramid *c.* 3000 BC
Some Egyptologists opine
that it was angled to align
on Thuban,* near the celestial Pole,   *α Draconis
whose starlight shone straight down the hole
for just a minute or two each day.
But it's impossible to say
if anyone intended this —   Since Khufu's corpse was laid
it's simply a hypothesis!   to rest,
                                                our wobbly axis has precessed.
                                                Though Thuban might be what
                                                    they saw,
Sexagesimals and star-groups   the two don't line up any more.
                                                (See p. 38)
The Tigris and Euphrates wind
across Iraq, and here you'll find
(turn to page 16 for the map)
Mesopotamia — the lap
where Europe's heritage was nursed.
Writing appeared in these parts first;   *c.* 3300 BC
and by a somewhat later date
the cleverest were numerate,
counting in *sixties*, if you please —   The sexagesimal system
hence seconds, minutes, and degrees.

Among their other innovations
were what we know as *constellations*.
Perhaps these were delineated
from Ziggurats, which punctuated
the level Mesopotamian plain.

2100–600 BC

A number of these sites remain –
stepped temples, with an outside stair,
but all in very bad repair.
A priestess on the topmost level,
all decked out for her nightly revel,
lay waiting for the local god;
but it would have been rather odd
if some star-gazing pioneer
did not climb to a lower tier
to take advantage of the height
and revel in the awesome sight.
He may have noticed, as he sat
observing on a Ziggurat:

'If I allow my mind to wander,
I see a scorpion up yonder!'
[Or eagle, snake, or what you will.]
The constellations grew, until
when we arrive at Homer's date   *c.* 700 BC
their number comes to 48.

Meanwhile . . .

Europe had undergone a thaw
two or three thousand years before;
and stones in rows and circles prove
that things were really on the move,
for they were hauled, at great expense,
to be set up as monuments.

Stonehenge is bound to come to mind,   *c.* 3000–1500 BC
It's been most carefully aligned

upon the first effulgent ray
that strikes it on Midsummer's Day;
but was it really meant to be
(as some say) an *observatory*?
And what about the other sites?
A venue for disgusting rites?
A place that only priests could enter?
An early form of shopping centre?
The truth about these works of stone
is likely to remain unknown . . .

Recycling

Each day the Sun gives heat and light
(though who knows where it goes at night),
and after every thirty days
the useful Moon repeats its phase.
It's true that star groups disappear
at certain seasons of the year;
but that's for just a month or two —
they always come back into view!
These cycles in the world at large
suggested someone was in charge . . .

Months or years?

It was the Moon and not the Sun
that they considered No. 1.
It governed birth, growth, and decay;
it waxed, grew full, and waned away;
so naturally they watched and cheered
when in due course it reappeared —
seeing the New Moon in the west
would set their anxious minds at rest!
The lunar cycle or 'lunation'
(its period of circulation)
became our month, although the days
no longer match the lunar phase.
If every month began with New,
which sounds a pleasing thing to do,
there'd be 11 days to lose
when New Year came (although some Jews,
and Muslims everywhere, prefer
to use a lunar calendar).

Cause and effect

It's not surprising if they felt
that some controlling power dwelt
within the Moon, and kept it going.
What of the planets' to-and-froing?
Their movements caused immense concern . . .
they'd go one way, slow down, then turn
and put themselves into reverse,   Retrograde motion
and then go on again. Still worse,
you might get two or three in line,
which had to be some sort of sign.
If any *comet* came along
things would most certainly go wrong;
and who knows what might come about
when Sun or Moon was blotted out?
But now, with Greece's golden age,
we enter an exciting stage . . .

2 Lively Minds
(c. 700 BC–AD 200)

An enterprising race

The denizens of the Aegean
were previously Mycenaean
(the *Iliad* and *Odyssey*
profess to tell their history).
Then came the Greeks from further north,
whose genius would usher forth
our present age. This vibrant race
took up a vast amount of space
from what's now Turkey in the east
as far as Italy, at least.
What's startling is their confidence
that by applying common sense
(with lots of geometry thrown in)
they'd suss it all out. Let's begin . . .

c. 1500–1100 BC

c. 800 BC

Thales gets wet and predicts an eclipse   c. 625–545 BC

In Thales of Miletus, we find
a foretaste of the Lively Mind.
While looking at the stars, he fell
(folklore assures us) down a well . . .
perhaps that's why he came to think
that water isn't just to drink,
but is the Primal Entity
that fashioned everything we see?
When he said an eclipse was due,
and his prediction turned out true,*
Milesians were most impressed
(they didn't realize he had guessed,
using a secret rule of thumb
that said when an eclipse *might* come).

*In 585 BC

The Pythagoreans c. 550–400 BC

Pythagoras began a sect
that held both genders in respect:
they lived in contemplative quiet,
beans were omitted from their diet,
and they believed that souls migrated.
One member, Philolaus, stated   c. 480–405 BC
that there exists, beyond our sight,
a Fire burning day and night,

the Sun reflecting back its rays,
and everything goes round the blaze
(he also managed to throw in
a second Earth, our ghostly twin).
But credit give where credit's due:
he was the first to take the view
that we're not fixed, but on the move —
which took two thousand years to prove!

Although this theory seems bizarre,
his interest in what things *are*
shows how a passion for reflection
pushed Mankind in a new direction.

Pythagoras and his triangle

c. 560–480 BC

Pythagoras thought the world a sphere
(a new and startling idea!),
but his enduring claim to fame
is through the Proof that bears his name.
Draw any triangle you dare,
one corner being nice and square:
the shorter sides call a and b,
the long hypotenuse call c . . .
then c^2 equals, you will find,
the squares of a and b combined.
A mighty intellectual stride
to make a statement that applied
to triangles as yet undrawn!
This moment seems to mark the dawn
of Reasoning Investigation,
which was of fairly short duration,

[27]

because, as I explain below,
what can mere mortals really *know*?

Plato and the problem of Perfection *c.* 427–347 BC

But problems followed in its wake . . .
It is impossible to take
a pen or pencil, and to draw
a triangle without a flaw:
no right angle is ever Right;
no corners tidily unite;
the Theorem, anyway, defines
a shape whose sides are perfect lines
(no width, and absolutely straight).
This figure one can *contemplate*;
but it exists as an Ideal
within the mind, not something real.
So artefacts are a delusion!
Thus, Plato came to the conclusion
that all the things we seem to see
are Shadows of Reality.
Perception's flawed, so *think* things out –
the Mind alone is free from doubt.
This principle proved heaven-sent
when faced with that experiment
in Pisa . . . Galileo's weights   See p. 132
ought to have dropped at different rates,

which they did not *appear* to do —
so Plato's principle was true!

Eudoxus models the Cosmos

c. 408–347 BC

Plato began a Learning Centre.

The Academy

Only the brightest males could enter:
they mused beneath the olive trees,
and most of them got Ph.D.s.
Eudoxus was a graduate.
He thought he'd try to predicate
the planets' movements, by creating
a *model* . . . thus initiating
a task that took two thousand years!
He nested 27 spheres
in seven groups (some four, some three),
all interlinked harmoniously,

[29]

and centred on the Earth — each one
controlled the planets, Moon or Sun.
Each planet's dedicated group
produced its 'retrograding' loop
when it slowed down, paused and reversed
(though in the case of Mars, the worst,
Eudoxus finally admitted
that none of his solutions fitted).
Beyond these tricky combinations,
one vast sphere held the constellations . . .
In line with what his Master taught,
it's doubtful if Eudoxus thought
that this was how the Cosmos ran —
such knowledge was beyond mere Man!

It simply helped him to *compute*.
A second student followed suit . . .

Aristotle and transparent spheres

384–322 BC

It is through no fault of his own
that Aristotle's widely known
as someone who held progress back.
His name spearheaded the attack
on Galileo, which implied
that Aristotle had denied
the value of collecting data.
It was revisionists, much later,
who thus besmirched his noble name,
and for that, he is not to blame –
in fact, he loved to poke around
and document the things he found!
Bees, fish and molluscs – anything
that went by water, land or wing –
received most scrupulous attention.
(In this connection, I should mention
that he had tutored Alexander,
the charismatic Greek commander,
when he was Small; when he was Great,
he sent back samples by the crate
from places he was conquering.)
But what did Aristotle bring

Alexander the Great 356–323 BC

*What Aristotle thought
and taught (in short)*

1 Beyond the Moon

The objects that we're privileged to see,
turn round the Earth, and represent Perfection.
The planets move in circles, constantly –
they only **seem** to alter their direction!

Recycling was God's method – not Creating.
He got hold of some quintessential stuff,
made crystal spheres, and set them all rotating.
A gentle push from him was quite enough . . .

2 Beneath the Moon

No changeless plan, no perfect circles here –
all's transient, chaotic, in decay.
Motion is *straight* in the sublunar sphere –
the elements prefer to move this way . . .

Fire, which rises; *Earth*, which falls instead;
Water and *Air*, which flow or blow along.
These are some things that Aristotle said.
It took two thousand years to prove Him wrong!

to our concern — Cosmology?
A lot, as we shall later see!
He liked Eudoxus's ideas,
doubled the quantity of spheres,

To 55 altogether

made them transparent, and insisted
that they were *real*. They existed,
he claimed, in an unchanging state

*This policy of fixity
would cause great problems, as
we'll see.*

(to human eyes, at any rate).
The only changes happened here —
on earth, beneath the lunar sphere . . .

Aristarchus decides to move the Earth

c. 1310—230 BC

The concept of the city state
was fast becoming out of date:
a conquering Hellenic Nation
was Alexander's aspiration.
He was, in fact, already dead
when cheeky Aristarchus said
that we go round the Sun. This notion,
which set the solid Earth in motion,

implied that it was also *spinning*
to give each day a new beginning.
Absurd! A gale would be blowing
against the way that we are going!

He also tries to measure the Sun . . .

Another thing he tried to do
(and this was most presumptuous too)
was find the distance of the Sun.
His method is a brilliant one,
a quantum leap in human thought —
so *concentrate* (I'll keep it short).

. . . using this method

By now there was no doubt at all
that, like the Earth, the Moon's a ball,
and the reflected solar rays
are what produce the lunar phase.
Now, when just half the Moon is lit
(this is the really clever bit)
the angle Earth-Moon-Sun subtends
90°, on which depends
his method. Having watched and waited
till it was half illuminated,

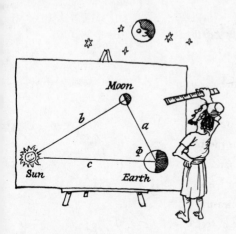

he checked the angle in the sky
between the Moon and Sun (called Phi);   Greek φ
and once he'd found it, he produced
a Triangle, and so deduced
the distance that the Sun must be
(in other words, the long side c)
compared with the much shorter a
(which shows how far the Moon's away).
His value for the ratio
was 25; and we now know
our warm and beneficial star
is some 400 times as far
as our near Neighbour of the Night.
But guessing when the phase is right
is very hard . . . give it a try,
and tell me what *you* get for Phi!

The publishers will redirect
your postcard, e-mail, or letter.
Don't worry if you're not correct —
just see if you can do much better!

Eratosthenes measures the Earth ...   c. 276–195 BC

To Egypt Alexander came
and built a city in his name,
its famous Library exceeding   Alexandria founded 332 BC
all others in its choice of reading.
Though Eratosthenes was meant
to manage this establishment,
he found he had sufficient leisure
to carry out his plan to measure
the Earth's circumference in feet.
His method was extremely neat.
If you peer down a well at noon
around the 21st of June

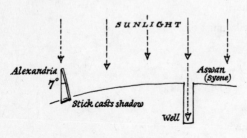

at Aswan, then known as Syene,
the bottom's sunlit, which must mean
the Sun's exactly overhead.
This simple observation led
perceptive Eratosthenes
to note that it passed 7°

south of the zenith where he was.
This happened, as he knew, because
the Earth is round; and so he got
some 'steppers', who were paid to trot
exactly, and to count the paces
that separated these two places.
The value of the Standard Stride
was now most rigorously applied;
the total distance was divided
by 7; and having thus decided
upon the length of one degree,
the rest was trigonometry!
(They must have trotted quite precisely –
the answer came out very nicely.)

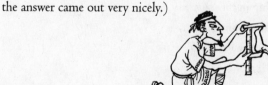

Hipparchus discovers Precession
c. 190–125 BC

Hipparchus was extremely good
at making gadgets out of wood
to measure the celestial place
of stars (he also liked to trace
the planets' movements – as we'll see,
these were a major mystery).
When he compared his observations
with those of earlier generations,
he realized that the stars had drifted,
but all together – they had shifted
by something over 1°
from right to left each century.
Our axis slowly twists around –
the wobble that Hipparchus found
is called *precession*, and that's why
his observations of the sky
secured his lasting reputation.
His legacy of observation
using his graduated arc
(not easy, working in the dark),
was used by Ptolemy, much later,
to make his Planet Calculator!

The wobble is extremely slow.
26,000 years or so
must pass till our aberrant spinning
returns to make a fresh beginning.

Ptolemy: the last Alexandrian   *c. 70–160*

Rome flourished, in its greatest years,
by raising civil engineers –
their bridges always took the weight;
their roads were scrupulously straight;
their aqueducts were watertight;
their sewers were a sheer delight.
From infancy their youth was taught
to waste no time on Abstract Thought;
so, as the Empire extended,
most non-productive thinking ended.
But Alexandria kept on
(although its Library was gone –
burned down by Julius, it's said,
while sharing Cleopatra's bed),
producing a stupendous Greek
whose contribution is unique.
To Ptolemy is due the plan
that would obsess the mind of Man
for fourteen centuries or so –
it's clearly something you should know . . .

Mark Antony, who loved her later
(though Rome considered him a
    traitor),
was so bedazzled by her looks,
he got her a fresh stock of books.

The puzzling planets

The planets' motions are the key
to Ptolemy's Cosmology.

They follow their appointed track
across the stars, then pause, go back,
then pause again and turn once more,
and go on as they did before.
How to predict this to-and-froing
and give ourselves a chance of knowing
the obstacles that lie ahead
(for in their motion, Fate is read)?
The *epicycle* was the key

This source of worry and
        confusion
is just an optical illusion,
produced, as you will later see,
by Relative Velocity.

to working out where they would be.
He started off with just a few,
and ended up with 32.

The *Almagest, c.* 150

The Clockwork Universe

How could a system so contrived
have been invented – and survived?
The reason is that (strange to tell)
it forecast everything quite well;
and Ptolemy would never claim
that *knowledge* was his final aim.
He thought we had no way of knowing
what keeps the stars and planets going . . .
all we can ever do is try
to *simulate* the changing sky,
using the skills that we possess.
In this, he was a great success!

Epicycles

Each planet moves at constant rate;
a *circle* is its course;
from this they must not deviate –
it is ordained, perforce!

This was believed to be the Law
that bound the heavenly sphere;
but if you watch them night by night,
discrepancies appear!

The wretched things accelerate,
or go into reverse.
The challenge to accommodate
behaviour so perverse . . .

. . . led Ptolemaeus* to invent
some cunning *combinations*.
Circles were linked, to circumvent
celestial regulations!

He used them like so many hoops
until they met his need
to reproduce the planets' loops
and regulate their speed!

These *epicycles*, as they're termed,
would doggedly persist,
till Kepler finally confirmed
that circles don't exist . . .

* Dear reader, do not be surprised
that Ptolemy's been Latinized.
Until Sir Isaac Newton's day,
names could be rendered either way.

More on *p.* 107

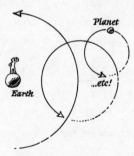

3 Staying Still . . .
(200–1200)

Two periods of history
end and begin with Ptolemy.
He would have been quite flabbergasted
to know how long his system lasted,
for up until the Reformation
all planetary calculation
was based on his ingenious scheme!
Why did it reign so long supreme,
when to our eyes it seems absurd?
It's partly due to what occurred
at Golgotha, at 3 p.m.,
one Friday in Jerusalem . . .

John's Gospel says the day before –
a fact most celebrants ignore.

Heavenly credit

The impact of what Jesus taught
pervaded all ensuing thought.

[43]

The most important thing was this:
that death will lead to heavenly bliss
for all who treat their neighbours well –
but if they don't, they'll fry in Hell.
So life is just a preparation
for people of whatever station –
in fact, the less you have, the better!
To keep Christ's teaching to the letter
means, first of all, *humility*.
Thought was exchanged for piety . . .
To question *why* a thing is so
is further than you ought to go,
for God arranged the cosmos thus –
to seek to know is blasphemous!

Burning the books

Rome, on the whole, had tolerated
the ways of those they subjugated;
but Christians were *evangelizing*,
so it is not all that surprising
that it became a public treat
when lions got a few to eat.

Especially under Diocletian, r. 284–305.

But their persuasion grew and grew;
and Constantine, the Emperor, knew

r. 306–337

that to avoid a major split
he really had to bow to it.

The consequences were far-reaching
when he embraced the Christian teaching.
It was enforced throughout the land:
the pagan gods were henceforth banned,
which meant that libraries were stripped
of every suspect manuscript.
Since Alexandria possessed
the greatest archive in the West,
the whole lot vanished in the grip
of prophylactic censorship.   *Museum destroyed in 391*

Augustine's cosmology

To set the medieval scene
I'll start off with St Augustine,   *d. 430*
whose book of Christian admonitions
went into numerous editions . . .   *Confessions, 397–8*

Corruption and predestination . . .

The choice that Eve was moved to make
when tempted by the subtle Snake
to sample the Forbidden Fruit
(and Adam quickly followed suit)
has left us in our sorry plight.
With that first catastrophic bite,

their heirs (i.e. the human race)
became dependent on God's grace
to go to Heaven when they died.
At this point, Augustine applied
the doctrine of *predestination*,
which lingered till the Reformation –
that most of us are for the Pit,
and doing good won't help a bit!
Augustine sensibly advised
that everyone should be baptized,
or even if they'd made the cut
they'd find the Gate of Heaven shut.
This doctrine went for infants too.
Few modern Christians share his view,
but babies are still robed and blessed
to set their parents' minds at rest.

This utterly appalling thought
is not a thing that Jesus taught.

The Creation . . .

Augustine also had a look
at what comes later in this book:
how was the Universe created?
Plato and Aristotle stated
that when God made it, He employed
whatever floated in the void.
This turned the Biblical Creator
into a mere facilitator,

which no believer could accept.
But here our saintly author stepped
on to more questionable ground . . .
if God has always been around,
it means that time must have begun
before He fired the starting gun.
Augustine struck a modern note
by stating, in the book he wrote,
that time did not begin until
God started to impose His will
by ordering: 'Let there be light!'
Cosmologists believe he's right,
though finding out *why* we exist
still baffles the materialist,
who tries to adumbrate the laws
that by themselves explain the Cause.

The 'Theory of Everything'

The age of the world . . .

He wondered, too, how long ago
the Universe began – although
to ask this question seems to me
to verge on impropriety.

Six thousand years was his suggestion;
an age accepted without question
until the record of the rocks
began producing seismic shocks,
and dinosaurs first came to light.   See p. 167
But, having no great faith in *sight*,
Augustine might not have been thrown
by fossils. 'Nothing can be known,'
he would have said, 'by mere *perceiving*.
Seeing is by no means believing!

Reason alone is free from doubt —
the Mind can work all problems out,
using the Scriptures as a guide.'
Well, for a thousand years they tried . . .

And yet he thought the Earth was round,
for which no sanction can be found
in Holy Writ. Roundness, and motion,
were held to be a pagan notion.

Dark times for Europe . . .

The Roman empire was falling —
the consequences were appalling.
Europe had muttered about taxes,
but wild men with spears and axes
were worse . . . they pillaged what they
    found,
and made the place a battleground.
The European light went out
with these barbarians about;
and in this Dark Age, as you'll guess,
Cosmology did not progress.

Rome sacked 410

Goths, Vandals, etc.

. . . but a light in the East

With Europe on its downward course,
Arabia produced a force
to rival it. Muhammad* heard
the Angel Gabriel's whispered word,
which he remembered and dictated.
His book (the Koran) clearly stated

*c. 575–632

[49]

that Christ was human, not divine,
though high in the prophetic line —
God's realm on Earth did not require
a figure known as the Messiah!
Obedient to its sacred text,
the faith of Islam soon annexed
North Africa, the Middle East,
and Spain; their gains might have increased,
but Charles Martel kept them at bay
in southern France, near Poitiers.   In 732
A hundred years was all it took
to bring this vast expanse to book.

Saving the archive

The rulers of these new estates
(or more correctly, Caliphates),
beheaded victims by the score,
and that's what they're remembered for;
but thanks to them the flame of learning,
which died in Europe, kept on burning.
The pagan works the Church had banned
(for reasons you now understand)
did not offend Islamic minds,
and so these almost priceless finds
were very carefully translated;
as Al-Jahiz of Basra stated,   'The Pop-eyed', 776–869

These ancient writings are the key
to Wisdom (that's a summary).
Unlike the likes of Augustine,
the Muslims were extremely keen
on Wisdom — which is not a word
that we have previously heard
in Christian mouths. Later, of course,
the West bewailed the resource
they had themselves so lightly burned,
and from which so much might be learned.
By then, the volumes they would need
were in a script they couldn't read . . .

Some of the words they used
    would stick,
for we still use the Arabic.
Ptolemy's *Syntax* is known best
by its new name, the *Almagest*.

Counting

Cosmologists are always counting!
As new discoveries keep mounting,
and distance records are exceeded,
still greater numbers will be needed.
The Roman system seems to us
so utterly ridiculous —
they'd write this year MMVI,   1000 + 1000 + 5 + 1
and I would like to see them try
to write one million using 'M',
the largest symbol known to them.
The Hindus, on the other hand,
used numbers we can understand,

i.e., the digits 1–9
(the Zero they did not define) –
and clever Al-Khwarizmi saw   *c.* 780–850
that this was what they'd waited for . . .

o is not nothing

Until then they had worked with beads,
which also met the Romans' needs;
but Al-Khwarizmi thought of *zero,*
which makes him something of a hero,
for zero is a multiplier
that makes a number ten times higher,   3 . . . 30 . . .
and if you add one more again
you once more multiply by ten,   30 . . . 300 . . .
and you could carry on all night
appending zeroes to the right!   300 . . . .
It was a turning-point in thought
to see what you could do with o.

The House of Wisdom

He flourished in Baghdad, now earning
top place in all the world for Learning.
Founded in 762,
the Muslims' eastern city grew
into a cultural oasis.
The House of Wisdom was its basis,
founded by Caliph Al-Mamun,   786–833
a learned man; and very soon
his store of volumes was immense.
He also built great instruments
some ten or fifteen metres high   Since they've been lost, I must confess
to measure objects in the sky,   that these dimensions are a guess.
and Al-Khwarizmi working nights,
kept all the planets in his sights,
comparing them with Ptolemy —
quite close, as far as he could see,
although his system was absurd!
Like Aristotle, he preferred
to think in terms of crystal spheres.
The same enthusiasm appears
(as we shall very shortly find)
within the hopeful Christian mind,
when post-Dark Age reflection starts
to fill space up with moving parts . . .

Our hero made one bad decision
and went for calendar revision.
Their year, which used the lunar phase,
was too short by 11 days,
so he said: 'Use the solar year,
and these odd days will disappear!'
However, since they liked the Moon,
this went down like a lead balloon.
Unfazed, he caused another stir
by coming up with Al-gebr,
i.e. *equations*, which would lead
to great conceptual leaps indeed,
as will appear when I start talking
of Newton and of Stephen Hawking . . .

The first New Moon seen in the sk
after the 16th of July
was the Islamic New Year's Day –
it has remained so, by the way.

Measuring the world (again)

Remember Eratosthenes?
He found the distance in degrees
between two points, stepped it, and hence
derived the Earth's circumference.
But Caliph Al-Mamun was set
on having his own estimate,
to bring more honour on Baghdad.
The best astronomer he had
was Al-Farghani, famed indeed –
his book was an essential read

See *p. 36*

fl. 833–861

for people in that sort of line
right up to 1669!

New edition of *The Elements*

He took charge of the measuring crew,
and Al-Khwarizmi joined in too . . .
His helpers checked the Pole Star's height
(it hardly moves throughout the night)
when measured at the base location,
then watched its change of elevation
when walking north (whereon it rose)
or south (it fell, as you'd suppose).

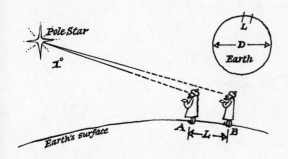

When the amount it rose or fell
reached one degree, the distance (L)
was measured — and as you can see,
the Earth's diameter, or D,
is $[(360/\pi)$
$\times L]$. It was a valiant try;
but what would Al-Mamun have thought
if he had known his D was short?

10,400 km, not 12,800 km

[55]

Naming the stars

About a thousand years before,
Hipparchus undertook the chore
of noting down the stars' positions. See *p.* 38
Though Ptolemy made some additions,
the list was getting rather dated;
and Al-Sufi is rightly fêted 903–986
for taking on the fearful slog
of going through the catalogue.
More than a thousand stars were there,
which he re-measured with great care.
Some of his names are still in use . . .
a well-known one is *Betelgeuse*, α Orionis
although the linguist much prefers
to call the same star *Betelgeuse*.

A shocking suggestion

The Omayyids had conquered Spain.
It's complicated to explain,
but from Muhammad there descended
two lines, whose feud is far from ended –
the Abbasids ruled in the east,
or held the reins of power, at least.
Toledo was the place to be
for students of astronomy . . .

here, Al-Zarkali (Latinized   d. 1087
as Arzachel) had just revised
the basic planetary data
and published his own calculator.   *Toletan Tables,* 1080
People were shocked, to hear him say
that circles were a bit *passé*
and it was time to get to grips
with modern shapes like the ellipse –

no one had ever dared imply
that circles didn't rule the sky.
(At least he held back from suggesting
the Earth should move, instead of resting.)

The break-in

Things started going wrong in Spain.
The caliphate was under strain,

and by the year 1031
a civil conflict had begun.
Although the Arabs' swift advance
had annexed parts of southern France,
four Spanish kingdoms had resisted,
and still precariously existed,
wedged up against the Pyrenees.
Alfonso reigned in two of these   Alfonso VI 1065–1109
(he ruled Castile and Leon),
and seeing what was going on,
he seized Toledo, setting free
the contents of its library.
That was the year, 1085,
when learning started to revive!

4 Wearing Spectacles
(1200–1500)

More scurrilous literature

The Universe, as we have seen,
was sorted out by Augustine . . .
St Ambrose (Bishop of Milan)   *c. 340–397*
agreed with him that sinful Man
should spend his time in preparation
for Life to Come – not speculation!
Popes earned great credit by Crusading –
they'd send their subject kings invading
the Holy Land (elsewhere as well),
to undermine the Infidel.
Constantinople had been sacked
in 1204 – the crowning act
of the triumphant Fourth Crusade,
which freed a positive cascade
of ancient and subversive stuff.
As if that wasn't bad enough,

it was *original*, i.e.,
in Ancient Greek, for all to see!
With these uncensored books on tap,
the Church was in a real flap,
and banned the works of Aristotle   Edict of 1209
a fellow they'd have liked to throttle
for saying that the Universe
was not created* — even worse,   *The matter had been there,
he said that we are much too small   unchanging —
for God to notice us at all!   God simply did some rearranging.
To think the troops of the Redeemer
brought back the works of this blasphemer!
However, as we'll shortly learn,
things took an unexpected turn . . .

Wealth in poverty

Our story, at this point, requires
some lines about the Begging Friars
or *Mendicants*, who gave up wealth
to boost their spiritual health.
In Matthew's gospel, Jesus said
'No need for money — you'll be fed!'   Matthew 10:9
when sending his disciples out;
these optimistic words, no doubt,
led to those Orders that were based
on being poor, sincere, and chaste.

St Francis set the trend, forsaking
his father, friends, and moneymaking;
St Dominic did so as well.
In one respect, their standards fell
(or rose) – for as their status grew,
Rome loaded in behind them too;
and as the show began to roll
they threw away the begging bowl.
Their cloisters were the place to be
for studying Cosmology . . .

Franciscans *c.* 1209

Dominicans 1215

St Francis went off to convert
a Sultan, and was rather hurt
when he eventually got back,
to find his team had built a shack.

Magnus and Aquinas ponder . . .

Two students from St Dominic's
came up with an inspired mix
of pagan thought and Christian lore –
just what the Church was praying for!
The Stagirite* had spent some years
arranging sets of crystal spheres
which nest within each other, thus . . .
(he ended up with 50+).
To some a planet is attached;
the movements of the rest are matched
so that each regulates its neighbour.
The outcome of this massive labour
was disappointing, I'm afraid,
because the planets quickly strayed . . .

*Since Aristotle's hard to rhyme,
I'll use his pseudonym this time.
He came from Stagira in Thrace.
Not much else happened in that place.

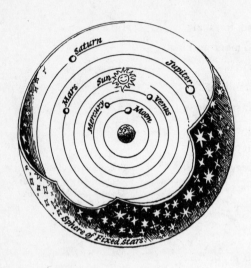

but spheres so cleverly designed
gladdened the medieval mind,
and got Albertus Magnus thinking

c. 1193–1280

there *had* to be a way of linking
this cosmic scheme with Holy Writ,
which ought to put the lid on it!
Thomas Aquinas, by and by,

c. 1225–1274

in *Summa Theologiae*

Principles of Belief, c. 1265

(as this defining work is known),
set it all out in words of stone . . .

His saintly stature is immense,
and had a lasting influence.
His views on birth control
    were clear –
it's not for us to interfere.

... and adopt (and adapt) Aristotle

To reach the goal that they were seeking
demanded some substantial tweaking.
First, Aristotle had asserted
that God, to make the world, *converted*
existing matter. Genesis
did not, of course, agree with this,
for His hard-working week began
with Nothing, and fetched up with Man!
And as for His indifference,
how could that possibly make sense,
since He had sent His Son to us?
The fallacy was obvious –
these things are there for us to see
to glorify His Majesty!

Angel power

Since Aristotle had professed
that matter likes to be at rest,
this meant it must be *pushed* along.
We're well aware that this is wrong,
since Newton earned his keep by showing
that matter simply goes on going;   That is, *momentum*
but, building on the 'push' idea,
the two designers gave each sphere

[63]

a workforce from the heavenly host
to turn it round. The outermost
(pronounce the last 'e', by the way)
is called the *Primum Mobile*,
which means the first revolving part –
God is beyond, just off the chart.
Next come the stars, which turn as one,
and then the planets and the Sun;
and closest to the sinful Earth
we find the sphere of lowest worth,
which bears the Moon. But that's not all . . .
for when the Damned are judged, they fall
into the Pit of Hell, and enter
the Devil's realm, right at the centre!

Hell – and beyond

Jesus (the Creed explains) *descended*
to Hell, after His Passion ended.
The Earth, all now agreed, is round,
so Satan's kingdom must be found
within the globe, beneath our feet.
This made the Cosmic plan complete,
since Heaven and Hell were sorted out!
But wait a minute . . . what about
the halfway house the Church created,
called Purgatory? That's located

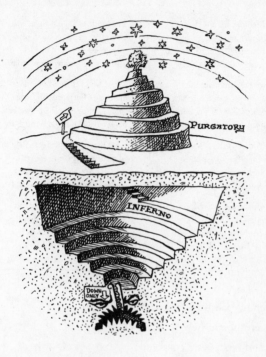

beyond the ocean, opposite
the part we know – the sunny bit –
down there, in antipodean gloom,
souls fidget in the waiting room.
In Dante's verse, the reader's told
what he and Beatrice behold
as they climb up the final stages
to Paradise (it takes them ages),
and find, upon the topmost tier,
that Eden's re-created here –

The Divine Comedy,
by Dante Alighieri, 1265–1321

[65]

Behold the Medieval Plan –
resplendent Order and Degree!
Above, the heavenly heights; then Man;
then Satan's realm, as you can see!

The human eye cannot make out
these crystal spheres, needless to say;
but they are there, beyond a doubt –
the Cosmos *must* be made this way!

The realm beyond the lunar sphere
is changeless, and has always been;
but in our murky atmosphere
all sorts of dreadful things are seen!

Nothing is safe beneath the Moon.
Those phantoms of the fiery air,
the *comets*, warn Mankind that soon
there'll be a spot of trouble there!

The inside of the Earth, as well,
forms part of the celestial plan . . .
the ever-open mouth of Hell
gapes wide for unrepentant Man.

Those who belong to the Elect
must enter Purgatory, and wait
while all their details are checked,
before they're ushered through the Gate!

Since prayers on Earth can help them then
(if their account is in the red),
the Church advises dying men
to fund some Masses to be said . . .

and so the chanting voices rise,
according to how much they'll pay;
and if it works, their future lies
beyond the *Primum Mobile!*

---

the launching-pad for Heaven! It's true
that this is a poetic view –
but fact and fancy were combined
within the medieval mind,
so there's no need to be surprised
at seeing concepts concretized . . .
if Purgatory goes up like that,
then draw it as a Ziggurat!

No need to look . . .

By 1300, then, we see
a confident Cosmology.

The synthesis had been perfected –
from Hell to Heaven, all connected!
The spheres were turned by angel forces,
so who cared if the planets' courses
wandered a bit from Ptolemaeus?
Such trivia should not dismay us!
Only Alfonso, dubbed 'The Wise',
showed any interest in the skies:
when printing was invented later,
his planetary calculator
remained the very best there was!
He's famous nowadays because
he wrote the first known book on chess –
so what's the yardstick of success?

Alfonso X, King of Leon
    and Castile 1252–84

Alfonsine Tables first printed 1483

The system Ptolemaeus used
left poor Alfonso quite bemused.
'If God had asked me, I'd have
    stated:
"Lord, *please* make things less
    complicated!"'

Reading the signs . . .

Astrology, as you'll have guessed,
was of absorbing interest.
The feeling was by now ingrained
that all disasters were ordained,
and that the Heavens mutely signed
what Providence had got in mind!
Without a doubt, the lunar phase
changed people in all sorts of ways,
as did its progress round the track
referred to as 'the Zodiac'.

The Church's attitude was mixed . . .
To say that everything was *fixed*
(i.e., mankind had no free will)
was not what Jesus taught; but still,
as long as nobody suggested
that stars and planets were invested
with power matching that of God,
it gave astrology the nod . . .

No GP would prescribe a cure
till he was absolutely sure
the signs supported the proceeding
(which usually came down to
  bleeding).
However tenuous the hope,
it helped to cast a horoscope.

Getting warmer . . .

But now and then, the Sky at Night
gave everyone a real fright –
the comet of 1264,
which lasted for two months or more,
made all who saw it so afraid
they simply rushed indoors and prayed.

Even Albertus Magnus wrote
that it was sensible to note
the planetary signs – e.g.,
Mars always meant hostility.

Comets meant famine, war, or plague.
The augury was often vague,
but hindsight (or a clever guess)
would prove their efficaciousness.

Comets, they thought, were very near,
which greatly added to their fear,
since Aristotle had insisted
that no such blemishes existed
beyond the Moon (God had arranged
that these parts would remain unchanged).
But what *were* these appalling things
that presaged dreadful happenings?
Well, fire rises, does it not?
This means the upper air is hot,
and so a comet's apparition
is caused by flammable emission
ascending, heating, and igniting.
Ingenious! But in the writing
of Robert Grosseteste, we now find
a leading medieval mind
trying to postulate a link
(for almost the first time, I think)
between a Cause and its Effect.
This was the start, in retrospect,
of practical investigation
instead of, well, indoctrination . . .

Chancellor of Oxford University,
Bishop of Lincoln, *c.* 1175–1253

He came from the Franciscan
     school,
which thought more freely, as a
     rule,
than the Dominicans would do.
They nurtured Roger Bacon too.

Grosseteste's prophylactic theory

He set himself to answer *why*
a comet shining in the sky
is bound to trigger something dire . . .
Suppose the essence that caught fire
had been of benefit to us,
composed of something virtuous –
a spiritual antidote
against corruption? As he wrote
in *De Cometis*,* its ascent
would leave our whole environment
exposed to every kind of evil.
That would account for the upheaval
that comets usually created . . .
until the loss was reinstated!

A 'virtue' was a power or force
sent by some quintessential source.
The Pole Star's virtuous emission
made iron point to its position.

On Comets, c. 1226

Out of perspective

What must the Bishop's friends have thought?
Enquiries of any sort
that questioned *why* things ought to be
were verging on impiety.
God was the Maker of the laws
that framed the Cosmos, and the cause
of everything that came to pass!
And now what? Taking bits of glass

that had been differently curved,
he noted down what he observed
when looking through them, even though
(as anybody ought to know)
these devious transparent slices
were really devilish devices
to make the world appear distorted!
Grosseteste eventually reported . . .
'Transparent media,' he said,
'can turn Perspective on its head.
Things near seem far — things far away
seem so close, they're as clear as day!'
The Bishop didn't have a clue
why lenses work the way they do;
but he suspected a connection
between their profile, or cross section,
and the effect that they produced.
From this, our pioneer deduced
that *natural* laws applied, and might
(with further research) come to light . . .

Ophthalmic remedies like these
had been employed by the Chinese
for fifteen hundred years at least —
but not much leaked out from
the East.

A lesser Leonardo?

His young disciple, Roger Bacon,
continued on the route he'd taken.
He's the more famous of the two,
acclaimed for things he didn't do . . .

1214–94

the Telescope, it's fairly clear,
cannot be reckoned his idea;
neither can Glasses for the nose,
or Gunpowder, which some suppose
to be this Oxford friar's invention!
I've only given him a mention
because you'd wonder where he'd gone
if I did not – so let's move on
to people who did more than he
to influence Cosmology . . .

The Order's high–ups took offence
at Bacon's strange experiments.
Suspecting him of heresy,
they kept him under lock and key
for fourteen years; but all in vain –
he came up with the Aeroplane.

Sharpening the razor

As international trade increased,
black rats, imported from the East,
spread infestation far and wide.

In the 1340s

A quarter of the people died,
including, so it is surmised,
William of Occam. He advised
simplicity in theorizing . . .

c. 1295–1349

and so it's not at all surprising
that spheres propelled by angels' wings
had no place in his scheme of things.
Keep theories to a minimum
is still a useful rule of thumb
for research in the present day
(called Occam's Razor, by the way).

The Occamists, a growing force,
began the gradual divorce
of Theory (led by observation)
and Doctrine (based on divination).

Adrift again . . .

The dismal fourteenth century
has no place in our history . . .
we'll let the swollen Earth digest
the fevered flesh on which it pressed,
and move on to 1401,
when Nicholas, a boatman's son,
was born at Cues, by the Moselle.
He ran away, and did so well
that he achieved prestigious rank
within the Church. We've him to thank
for stimulating fresh debate
on how the planets circulate.
He wrote these words in 1440
(doctrinally extremely naughty):
'The Cosmos isn't like a ball!
In fact, it has no shape at all,
and everything is on the move . . .
so how could anybody prove
the middle's here? It might be there . . .
in fact, it could be anywhere!

With no fixed centre, I submit,
the Earth *must* move around a bit!'

... *but not going round in circles* ...

To have a *Cardinal* suggest
the Earth is not, in fact, at rest,
must have been headline-grabbing stuff.

Though by the time his book appears,
he has been dead for fifty years.
(*Learned Ignorance*, printed in 1514.)

As if that wasn't bad enough,
he came up with this blasphemy:
'You can't have Circularity
without a centre.' Circles, too,
were out, if his premise was true!

... *and not judged!*

Here is the third and last sensation:
'How can there be a graduation
in terms of pre-appointed worth,
beginning with the central Earth

(this vale of tears, this moral slum),
to outermost Elysium —
if there's no centre and no edge?'
A real bombshell, to allege
the Cosmos has no master plan
to underline the state of Man!

Starting the great debate

So, helped by Nicholas of Cusa,
the medieval bonds grew looser,
though this was done by asking 'Why?'
instead of looking at the sky.
The Europeans didn't care
for stargazing — I'm not aware
of any medieval site
where they observed the sky at night.
(When King Alfonso's team updated
the *Almagest*, they just collated
antique Islamic measurements).
The Cardinal's accomplishments
included the initial training
of Georg Purbach — and he, on gaining
Vienna's Mathematics Chair,
arranged a public meeting there,
where delegates expressed their views
on whether they would rather choose

See p. 68

1423–61
This highly influential man
admired the Ptolemaic plan;
but, like the Cardinal, confessed
that sometimes an ellipse
   worked best.

a fixed earth, or an earth in motion . . .
This revolutionary notion
was therefore something to discuss
decades before Copernicus!

Opening the window

Just now I said, as an aside,
that Europeans had relied
on old Islamic observations.
Georg had the deepest reservations . . .
'It's time,' he thought, 'to make some more!'
Observing from an upper floor,
he and his brilliant protégé
(Johannes Müller) worked away,   1436–76
using contrivances they made,
and were exceedingly dismayed
to find Alfonso's *Tables* erred!

How could this be? What had occurred?

Soon afterwards came his demise;
but Müller, whom we Latinize
as Regiomontanus, took
a leaf out of his master's book.
With wealthy Bernard Walther's backing,   1430–1504
this prodigy began attacking
the basic problem — *lack of data*.
He thus became the joint creator

Mars was the most demanding test.
It strayed much further than the rest
because its path is far from round —
as Kepler (with much labour) found.

of Europe's first observatory,
at Nuremberg. Its inventory
mentioned a clock, whose heavy weight
turned wheels at the proper rate
to show the time (well, more or less*).
And what is more, a private press
(that's *two* surprising innovations)
was used to print their observations.

Founded 1471

*Crude mechanisms rang a chime
two hundred years before this time
but in their use as mentioned here,
Johannes was a pioneer.

If only . . .

Alas! Four years were all he spent
in this unique experiment.
The calendar was nine days out:
in Rome, the word had got about
that Regiomontanus* knew
(or ought to know) what they must do.
So off he went, and there he died,
the calendar unmodified.
It's said that someone did him in,
although the evidence is thin –
the plague is a more likely cause . . .
but had he lived, would the applause
for Canon Koppernigk (then three)
have been for him? Undoubtedly!
He had the very latest gear
to measure with; and it's quite clear

*He's known as this because he
came
from Kônigsberg, which means
the same
(the Hill or Mountain of the
King).
It has a most impressive ring.

Nicholas Copernicus 1473–1543

that he was firmly set on proving
that we aren't fixed at all, but moving.
And once he'd done the calculating,
his private printing press was waiting!
But it was not to be. Instead,
the one to turn things on their head
would be the fellow we all know;
but first, to Italy we'll go . . .

5 *No Fixed Abode*
(1500–1609)

The context

The Renaissance hit Italy —   *c.* 1400
a time of creativity
unmatched in grandeur ever since.
Each state or region had a Prince
who had the wherewithal to pay
the greatest masters of the day
to celebrate in paint or stone
the power centred on their throne!
Rulers, of course, had always spent
vast sums on self-aggrandizement —

what made things different, was the feeling
of *competition*. Paint a ceiling
(or wall), or spend years on a tomb,
or simply decorate a room . . .
Just make sure it's the very best,
and leave your visitors impressed!

Eyes on the ground

It couldn't last. Such sentiments
lead rapidly to decadence
when patrons start to ask for more
of what they know has worked before.
But what astonishing conviction . . .
the image of the Crucifixion,
the Bible legends, every saint
immortalized in stone or paint!
What walk-on part did Learning play
in this extravagant display?
The search for Wisdom that we found
when wealthy Caliphs were around   See Chapter 3
is very far from obvious.
It may seem rather odd to us
that neither Pope nor Potentate
(who must have known of the debate
about the Cosmos) did their bit
in trying to make sense of it!

An arc by Michelangelo
(still better, two) would surely show
that we were firmly in our place?
But no — they *knew* that was the case.
Why make a fuss? Damn those who doubt,
and Lucifer will sort them out!

Canon fodder

So, far below the Sistine vault,
Copernicus planned his assault

upon the notion underpinning
the Church's view, from the beginning —
the Earth is central and unmoved!
But how could Scripture be disproved?

It's possible he went to see
the Nuremberg observatory;
but he was not of Müller's mould.
Stargazing simply left him cold –
just 27 of his sightings
are mentioned in his published writings.
'Why bother?' he would have replied.
'With data I am well supplied.
The problem is digesting it,
and making all the planets fit!'

Planning his worstseller

Till 39 or thereabouts
he harboured his unpublished doubts.
He'd lost his parents as a boy,
and earned his keep in the employ
of Uncle Lucas, as Physician
and Secretary – a position
that gave him ample time to ponder.
When Uncle was escorted Yonder,
Nick moved to Frauenberg, and spent
three decades living free of rent,
without a great amount to do –
they gave him his own tower, too.
Here, in the great Cathedral's shade,
he dilly-dallied and delayed

Bishop of Ermland

He died in 1512

[83]

till, lying in his mortal bed,
he touched his book,* which no one read . . .

*De Revolutionibus Orbium Coelestium
(On the Rotation of the Heavenly
Spheres), published February 1543

Closing the loophole

Much earlier, say 1510,
our hero sharpened up his pen
and wrote an outline of his plan.

Commentariolus (Little Commentary)

This low-key manuscript began
with an astonishing admission –
he had demoted our position
to *strengthen* Aristotle's law!
There was a fundamental flaw
in Ptolemy, who'd done a fix
(see the adjacent limericks)
which contravened the regulation
forbidding any variation
of orbital velocity –
all planets must move *constantly*.
This major loophole would be closed,
using the method he proposed . . .

Ptolemy's fix (the Punctum Equans)

The great Aristotle decreed
that a planet should always proceed
at a uniform rate.
That was simple to state —
but the working was complex indeed!

Poor Ptolemy struggled in vain
to make uniform motion explain
why a planet is *there*
instead of elsewhere . . .
But at last, his magnificent brain

dreamed up an ingenious fiddle.
Take a point* that is *not* at the middle
of the circle it traces.
He made this the basis
for finally solving the riddle!

*The *Punctum Equans*
('Point of Equality')

Now, if you imagine you're here
(at the *Punctum Equans*), it is clear
that when it's at ζ
the planet *seems* fleeter
than at π, when it isn't so near.

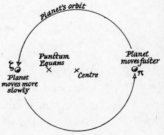

So he made its velocity high
when it's at its most distant, near π
and with great skill arranged
that its speed seems unchanged
wherever it is in the sky!

By doing all this, he could claim
that its motion was always the same
at all points on the curve
(which required some nerve);
so Copernicus made it his aim

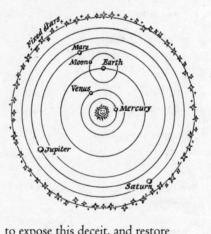

to expose this deceit, and restore
Aristotle's smooth motion once more.
In his view it was worth
unfixing the Earth
to maintain this unbreakable law!

Backing not lacking

The interesting fact is this –
the Canon's new hypothesis
offended hardly anyone.
On finding out what he had done,
the Pope's own Secretary led
a seminar, at which, it's said,

In the Vatican Gardens, 1533

the audience showed interest.
A Cardinal was so impressed,
he wrote: 'Dear Nick, your theory's great!
Publish it – don't procrastinate!'
How could the Papacy have backed
the most unpalatable fact
that calculations were improved
on the assumption that we moved?
Presumably, they took the view
that though the theory wasn't *true*,

Letter from Cardinal Schönberg,
1 November 1536

it did the calculations better,
and kept more closely to the letter
of Aristotle's laws of motion.
This utterly wrong-headed notion
was therefore something which, they thought,
deserved conditional support . . .

A decade of doubt

So it was Nicholas himself
who kept his writings on the shelf.
The text was finished (more or less) –
he'd written the complete MS
by 1530. It seems clear
that Doubt assailed him, not Fear.
Look at the price he'd had to pay
for perfect motion! Once a day

we're whirled around, but aren't sent flying?
Ridiculous! And then, denying
that we are in the central place —
how could he look God in the face?
He'd been quite certain all along
that Ptolemy's idea was wrong —
but though he'd done the best he could,
was his plan really any good?

Said Martin Luther: 'Who's this git
who's contravening Holy Writ?
Joshua stopped the Sun, thus proving
that on all normal days it's moving!'
(Joshua 10:13)

Priming the Canon

Joachim Retyk came to stay . . .
He was the *Primum Mobile*,
a young Sun-centred militant
(and, by the way, a Protestant,
who leaped the intervening chasm
of Lutheran iconoclasm).
He bullied Nicholas no end,
and in a few short weeks had penned
a meaty, bite-size distillation
and sent it off for publication.
But Nicholas would not permit
his own name to appear in it —
Joachim wrote *My Learned Master* . . .
However, things at last moved faster.
This 'First Account', the Canon knew,
implied that he must publish too —

Or Rheticus, 1514–76

From Wittenberg, already turning
into a five-star seat of learning.
Shakespeare has dropped a hint to us
that Hamlet was an alumnus.
(Act 1 Scene 2)

Narratio Prima, 1540

the world knew that his book existed!
The stubborn fellow still resisted,
but finally he gave permission
to print a limited edition.
So his amanuensis stayed,
revising all the text (unpaid),
which took a year or more, and found
a printer.* Once it had been bound,
Old Nick was shown one (so they say)
the very day he passed away.

*Joachim was extremely vexed
 to find no mention in the text
 of how he'd helped to get it
  finished.
 His rapture, after that,
  diminished.

The unread book

No other book, it can be said,
has caused more stir and been less read
than Nicholas's turgid tract.
What mattered was the simple fact
that it existed. People heard
(although they hadn't read a word)
about the radical idea,
whose implications were so clear
they didn't need to *buy* the thing!
Compare the people clamouring
to read what Darwin had to say —
his book's still selling well today.*
And as for helping with computing . . .
the plan he'd spent his life refuting

I hope this book sells more
  than his —
four reprints in four centuries.

* *The Origin of Species*, though,
  is gripping stuff — which goes
   to show
  that being boring helps your hand
  if you don't want your writing
   banned.

was just as good, and simpler too!
We must give Ptolemy his due –
counting the epicycles needed,
the old Earth-centred plan succeeded
with 40, whereas 48
were wanted in the up-to-date
Earth-moving scheme! And now's the time
(if I can manage it in rhyme)
to draw the reader's mind to what
would bother people quite a lot
about the moving-earth idea.
The box, I hope, will make this clear . . .

A waste of space . . .

With the Earth at the centre (as shown in Plan A),
the stars all remain the same distance away
on the outermost sphere as it daily turns round –
and throughout the whole year, the same patterns are found.

Earth

PLAN A

Plan B is, of course, the *Copernican* scheme,
where the Earth flies through space, and the Sun reigns supreme.
Let's call the Earth's orbital radius R.
Then our distance from any particular star
will change by as much as $2R$ every year . . .
So you'd think that the patterns of stars would appear
to expand and contract in a way that reflects
our approach and recession; but since these effects
are not noticed in practice, Nick's only defence
was to say that the stars' distance (S) is immense.
500 x R was his crazy suggestion
(which he later increased!). But this raised a big question . . .

Why leave all this space beyond Saturn? The fact
that the Earth-centred system is much more compact
was another good reason for people to wait
before casting their vote in the Cosmic Debate!

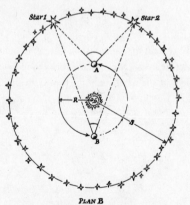

PLAN B

*The angle between Stars 1 and 2 should appear greater
when the earth is at A than when it is at B, unless
S is very large compared with R.*

A new thread . . .

With hindsight, 1572
saw Aristotle's scheme fall through.
A new star (which became so bright
that it cast shadows in the night)
attracted mystified attention.
Among its watchers, I must mention
the name of Thomas Digges, who found   1546–95
a piece of thread lying around,
and stretched it up across the sky
like this . . .

You may well wonder why!
Suppose the distance of the star
was not, in fact, so very far,
but that the others he could see
were almost at infinity.
In that case, then, its nightly drift
should show a *parallactic shift*

[93]

when it was near the east or west.
When Thomas tried his low-tech test,
no shift of any sort was seen.
This threw him, for it had to mean
that this prodigious source of light
lay somewhere in the heavenly height
beyond the lowly lunar sphere!
But no change was permitted here –
at least, so Aristotle said
(whose name was hanging by a thread).

The lunar parallax was known.
The work of Ptolemy had shown
a shift of roughly one degree
from where you'd think it ought to b

Introducing the Great Dane

As soon as the display was over,
a book appeared (*De Stella Nova*),
in which Tycho de Brahe proved
(like Digges) that it had never moved.
But Tycho did his measuring
with more than just a piece of string!
This sharp-eyed Danish noble planned
to outdo princely Samarkand,
where instruments in great array
(the pride and joy of Ulugh Bey)
made measurements that far outclassed
their predecessors. Now, at last,
a *European* monarch thought
he'd go for something of that sort . . .

The New Star, 1573

1546–1601

1394–1449

Cash for kudos

The Danish king fell in the drink.
His heavy robes caused him to sink,
so Tycho's father made a dive
and saved him – but did not survive.
King Frederick thus felt some blame
for orphaning young Tyge (his name
is now known in its Latin version);
and having heard of the coercion
from Switzerland, which tried to drain
the brain of this outstanding Dane,
the king got hold of him, to say
that if he would agree to stay
and let his skill in observation
add lustre to the Danish nation,
he'd offer him a choice of sites.
Uraniborg (the Heavenly Heights)
he built upon a rock he found
located in the Danish Sound . . .

Frederick II 1559–88

Splitting peas

Upon this island, known as Hveen,
the best observatory yet seen
was built at Frederick's expense.
What an array of instruments,

Commenced 1576

and what an unexampled eye
this fellow turned upon the sky!
If you behold a garden pea
at arm's length, it will seem to be
about the size the full Moon looks.
Now take the sharpest knife of cook's,
and chop the pea in fifteen slices –
our hero's cutting-edge devices
defined a star with a precision
equivalent to such division.

Invited foreign guests were sure
to add the island to their tour,
though what appealed to them most
was Tycho's standing as a host –
they cared less about matters stellar
than what he brought up from his cellar.

Including James VI of Scotland

Back to the centre

Uraniborg nursed Tycho's dream
of proving his 'Tychonic' scheme,
which was ignored by everyone.
The planets move around the Sun
(he thought), which in turn orbits us.
So he had dumped Copernicus!
The reason isn't hard to seek.
His accuracy was unique –

His plan is similar to Nick's
in its geometry. Unfix
the Earth, and centralize the Sun,
and it's just like the other one.

ten times as good, or maybe more,
as earlier work. And yet, *he saw
no parallax!* Try as he might,
his beady eye and knife-edge sight
discerned no side-to-side effect,
which he would certainly expect
the stars he had observed to show
if we were moving to and fro.

See p. 92

In Tycho's view, the choice was clear —
either the outer starry sphere
exceeds the compass of the brain,
or he must fix the Earth again!
And so he went for Fixity,
which suited his Cosmology —

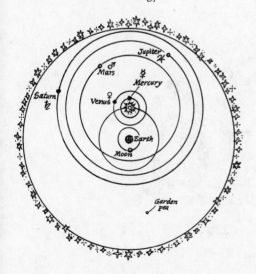

the central place regained by Man
exalted the Tychonic plan!

Hitting the road

Although he had survived his spill,
the good King Frederick fell ill
and died in 1588.
Things started to deteriorate,
for Tycho, as the saying goes,
had trodden on a lot of toes.
The honeymoon had been exotic,
but now his rule became despotic –
he had his tenants put in fetters,
refused to answer royal letters,

Of drink, his orator confessed,
when Frederick was laid to rest.

and then packed up and went abroad.
It's possible that he was bored –

In 1597

he'd catalogued a thousand stars,
made precious measurements of Mars
(which, as we'll very shortly see,
scuppered the old Cosmology);
and at the age of 50+
was starting to get curious
to see if all these observations
fitted his theory's expectations.
He spent two years hunting around
for patronage: at last he found
a suitably endowed position –
Imperial Mathematician
to Emperor Rudolph II.
In his new castle,* Tycho reckoned,
he'd sort his system out for good!
Alas, he died before he could,
and pleaded, lying in great pain:
'Don't let my life have been in vain.'
(*Ne frusta vixisse videar.*)
Relax – Johannes Kepler's here!

1552–1612

*He had a choice, and chose the nicer –
at Benatek, beside the Iser,
near Prague, the Emperor's domain.
But soon it was To Let again.

1571–1630

Two backgrounds

I should have mentioned, I suppose,
Tycho de Brahe's silver nose –
a form of surgical renewal
(he lost the first one in a duel,

a hazard of his social calling).
Kepler's upbringing was appalling.
His father ran away to sea;
his mother, tried for witchery,
was almost roasted at the stake
(what journeys he was forced to make
to argue in her desperate cause
while working on his famous Laws!).
Conceived before the bridal bed,
born early, hammers in his head,
his blistered epidermis flaking,
his wretched stomach always aching,
worms, piles, and gall-bladder trouble –
oh yes, and everything looked double.
To cope with this was rather worse
than sorting out the Universe.

Weil-der-Stadt, Germany

The Cosmic Cup

By good chance he was talent-spotted,
and took a place he'd been allotted
to study for the Church – these courses
were free, to help recruit the forces
the Protestant belief would need
if it was ever to succeed.
His life's recorded in his writing;
but it was rather too exciting –
religious hatred fanned a fire

Some 20 books have been
located –
I'm not sure if they're all
translated.

that polarized the Empire,
and soon all Europe was at war.

The Thirty Years War, 1618–48

But back to Cosmic things once more . . .
All students of theology
learned maths as part of their degree,
and obviously kept abreast
of what Cosmologists professed.
Kepler, who'd cut his studies short,
taught maths in Graz (a last resort);

1594–98

and on the 9th day of July
in 1595,* saw why

*While writing on the board that day,
the idea took his breath away.
He had a lot of work to do . . .
but that was the initial clue.

the planets number six, not eight
or ten, as he would demonstrate . . .
As Euclid long ago made clear,
five *perfect* solids fit a sphere
(all sides of each must be the same
for them to earn this hallowed name).
Kepler imagined all five nested
(you'll see the order he suggested
shown overleaf, so have a look).
Once he had ordered them, he took
a set of spheres, each sphere designed
with two adjacent shapes in mind,
so that it formed an interface
that held them in their mutual place.
The radii that he derived
for all the spheres he thus contrived,

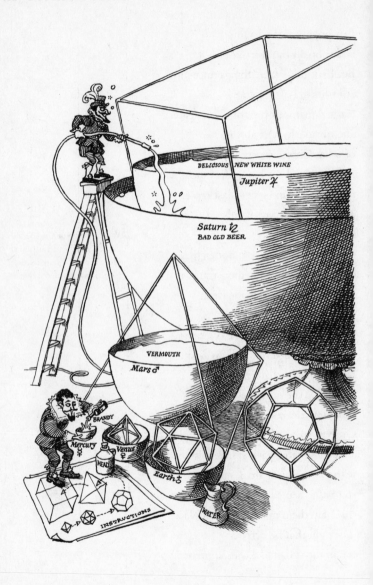

predicted (so he thought) how far
successive planets really are —
and Saturn's *got* to be the outer,
as he explained to any doubter,
because the shapes were all used up!
He planned a silver drinking cup,
a different beverage in each,
served by a tap in easy reach,
to let his Cosmical creation
toast its own cheerful celebration.
A Duke, on hearing Kepler's hype   Duke of Württemberg 1557–1608
(helped by a cardboard prototype)
wanted to have the cup created,
but it was much too complicated –
and also, struggle though he might,
the different distances weren't right.

God the mathematician

To modern minds, it does seem strange
that anybody could arrange
five polyhedra in a nest
and think the ratios expressed
some Cosmic principle or law.
Likewise, two thousand years before,
Pythagoras looked into Sound,
shortened vibrating strings, and found

that ratios like 4:3
produce a pleasing harmony,
but others grate upon the ear.   e.g. 6:7
He came up with the bright idea
that every planet sings or hums,
and their concerted sound becomes
a chord – the Music of the Spheres,
inaudible to earthly ears
unless in some ecstatic state.
Kepler himself would calculate
these cosmic sounds; but not before
he'd passed his First and Second Law . . .   See p. 107

Stormy weather

Stern faces under carmine hats
put paid to teaching maths in Graz.
The Lutherans were forced to flee
or fuel the fire of heresy,
so Kepler, who by this time guessed
that Tycho's measures were the best,
set off upon a lengthy trek   1 January 1600
to try his luck at Benatek.
It wasn't totally one-sided . . .   He'd left his family behind,
                                                 a worry always on his mind.
Tycho himself had now decided   His first wife went insane,
that Kepler's skills were what he'd need   and died –
to make his Cosmic Plan succeed.   his friends chose her to be
                                                   his bride.

But what a pair! The *grand seigneur*
assailed by the mangy cur

(that's Kepler's savage self-portrayal).
At first, it seemed to no avail –
Johannes asked him for the data,
and Tycho simply answered: 'Later!'
Before long, he had had enough,
left Tycho's castle in a huff,
and wrote him an abusive letter,
which didn't make things any better.
Tycho agreed to his conditions,
and let him have the Mars positions . . .
but deep inside, was he afraid
that all the sightings he had made
might undermine his great idea
('*Ne frusta vixisse videar*')
once Kepler got to work? Maybe.
Death saved that painful irony . . .

Referring to himself, he writes:
He's like a dog, and sometimes bites.

Data transfer

Before the funeral orations,
Kepler grabbed Tycho's observations
and took the whole lot 'into care'
(his words) in case the son and heir
disposed of them, or just refused
to let the Great Dane's work be used.
He also got his job, which meant
more money earned,* but much more spent.
The Emperor's demands were small
(his horoscope was really all);
but even so, it took him ages
to go through all the painful stages
described in *New Astronomy*.
He stated: *Chance discovery*
(at least as much as common sense)
guided my own accomplishments . . .
The title page (in upper case)
states that computing MARS's place
had been the challenge. Kepler knew,
as did his predecessors too,
that Mars is hardest to contain —
it simply wanders off again,
refusing to obey prediction.
Hence Kepler's deep-seated conviction

Buried in Prague
4 November 1601

*Though less than may at first
  appear —
  his pay was always in arrear.

Astronomia Nova, 1609

Kepler's first two Laws.

The fundamental Law (that's No. 1)
states that each planet's path around the Sun
is not a circle, as was always thought,
but an *ellipse*. A loop of thread, kept taut
against two pins (the foci), will produce
this pleasing curve, unless a pin comes loose.
One focus indicates the Sun's location.
Having defined the basic situation,
he tried to find how fast each planet went
at different points along its path's extent
(by methods that were largely hit-and-miss).
The answer was Law 2, which goes like this:
The areas of portions A and B,
swept out in equal periods, agree!

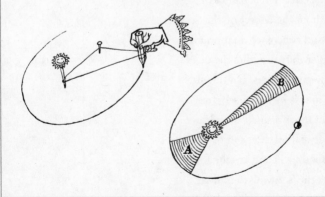

that Mars contained the vital clue.
The first thing that he had to do
was sort through all his pilfered treasures
for Tycho's finest Martian measures . . .
but he could *not* make them concur
with motion that was circular.
This must have been a shattering blow —
as far back as you cared to go
the circle seemed the Cosmic curve!
But since the circle wouldn't serve,
he had to make the best of it
and see if an *ellipse* would fit . . .

Help!

Ellipses worked — but what a cost!
The perfect circle had been lost!
Without the Mover of the Spheres,
the old coherence disappears.
Kepler thus straddles the divide
between two ages. On one side
his foot is kicking itself free
from the ingrained complexity
of crystal orbs kept in their courses
by God's will and angelic forces;
the other's waving in mid-air,
no certain footing anywhere!

Finding a force

The challenge that he had to face,
to keep the planets in their place,
was dreaming up some new connection
that moves them in the right direction . . .
If they go round the Sun, he mused,
the power that is being used
to keep them firmly on their course
has got to be a *solar* force!
Out of the blue, he postulates
that, like the Earth, the Sun rotates –
a fact that wasn't known till later.   Revealed by Galileo 1610
Extending from the Sun's equator,
a circulating force (perhaps
magnetically-based) entraps
the planets, sweeping them around.
The implications were profound . . .

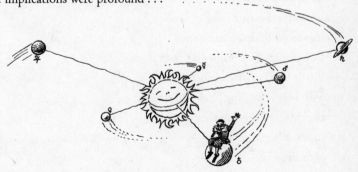

The tenets of theology
were based on motion by Decree
via the *Primum Mobile*;
but Kepler's model did away
with Will and an external cause —
implied, in fact, that earthly laws
are even true beyond the Moon!
It went down like a lead balloon,
and got him very bad reviews . . .
ellipses, of all things to choose!

Three decades before Newton's
   birth,
Kepler suggested that the Earth
attracts things to it; that the sea
is tugged by lunar gravity;
and that two masses, free in
   space,
would pull each to the other's
   place.
Newton did not refer to this
when working on his synthesis.

The Third (Harmonic) Law

It took him a decade or more
to come up with his final Law.
It's found in a miscellany
entitled *Worldly Harmony*.
This work includes his new ideas
about the Music of the Spheres
(based on Pythagoras's notion
that planets hum while they're in motion).
He thought of linking orbit size
with note — i.e., the pitch will rise
the closer to the Sun they get.
In his celestial septet,
Mercury's alto, Saturn's bass!
The Earth, such an unhappy place,

Harmonice Mundi 1618

See *p.* 103

The Moon makes seven

gives out a constant mournful wail —
the third and fourth notes on the scale.
They're Mi (for *Mi*sery) and Fa
(for *Fa*mine), which conflict and jar.
These esoteric explorations
included, in their revelations,
the Third Law, which I'll give you here . . .
The square of P *(a planet's year),*
divided by a³ *(its distance),*
is constant. Let me give an instance —

The Earth's speed slows, and then
it gains,
as Kepler's Second Law explains.
This variation in our rate
causes the note to modulate.

How to calculate Neptune's year in terms of the Earth's year, using Kepler's Third Law

Neptune's distance (a) is approximately
30 times the Earth's distance.
$a^3 = 30^3 = 27000$.
Neptune's P is therefore approximately
$\sqrt{27000} = 164$ times the Earth's year.
Using a precise ratio of 30.17 for a, the answer comes
out very close to the true value of 165.7 Earth years.
Try it!

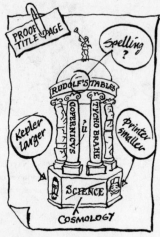

One last task . . .

Rudolph, by now, had been unseated,
but Kepler faithfully completed
the Tables that perpetuate
his patron's name. They tabulate
the Moon's and planets' future places
using his Laws (built on the basis
of Tycho's unexampled vision)
with hitherto unmatched precision.
The Frontispiece, his own, suggests
that old Cosmology, which rests
upon the pillars of the past
has started to collapse at last –
but two new columns take the weight,

By his brother in 1611

Rudolphine Tables 1627

which on their bases clearly state
COPERNICUS, and TYCHO! Where
is Kepler, you may ask? Not there
among the greats who steal the show —
you'll have to look for him below,
beneath the platform, out of sight
just like the printers (to the right)
who slave away behind the scenes
and get no thanks — that's what it means!
He'd trudged round treasurers, and pleaded
to get the back pay that he needed
to have the copy typeset, checked,
printed, and bound — so, in effect,
he was the volume's sole creator.
He died of fever three years later;
in his delirium, it's said,
he kept on pointing at his head
and at the sky . . . His grave is gone
(he passed away at Ratisbon);
but his memorial's enshrined
in those three Laws that he divined!

6 *The Fall of the Apple*
(1608–87)

Galileo's bestseller

Galileo Galilei 1564–1642

Now I must back-track, and relate
what happened in 1608 . . .
Holland was fighting to be free
of hated Spanish tyranny:
a Dutchman called Hans Lippershey
offered the State a magic eye
that made things far away seem near.*
Its military use was clear:
Lippershey pocketed the loot,
but others also pressed their suit,
and very soon the leisured classes
were snapping up 'perspective glasses'
or 'optick tubes' as novel gifts.
Two years pass, and the action shifts
to Tuscany, where news is brought
to Galileo . . . After much thought

*Though making nearby
objects vanish
was what they wanted for the
Spanish.

(or so he tells us) he deduced
how it must work, and soon produced,
through great exertion and expense,
a copy of these instruments . . .
Its revelations made him keen
to let the world know what he'd seen –
the book he hustled through the press
became an overnight success . . .

Sidereus Nuncius, 1610

I'm a Florentine Gent – Galileo's my name
(a Professor of Maths, by the way).
I am eager to share
what I've spotted up there,
so I hope you'll believe what I say . . .

The monster device I decided to point
at the Heavens, was made by myself.
It's decidedly stronger,
and its tube is far longer,
than anything bought off the shelf!

'But what have you *seen?*' I can hear you exclaim.
Well, the Moon is exceedingly rough.
Its peaks are so high
they would break through the sky!
And if *that's* not exciting enough . . .

. . . the stars in themselves are as small as before
(i.e., points) – but how many I've seen!
They're so greatly increased
I spot five more at least
where just one star was formerly seen.

As you would expect, I have saved till the last
the *pièce de resistance.* Hold tight!
Four new worlds rotate
in celestial state
around Jupiter, moving each night!

I think it is sensible not to provoke
the passions of those who deny
that the Earth's on the move –
yet Jove's satellites prove
that not everything up in the sky . . .

. . . revolves around *us!* Well, I'll go on my knees
(in due course) at the famed Inquisition,
when I'm forced to agree
(despite what I see)
that the Earth doesn't change its position!

Meanwhile, Grand Duke, please accept as a gift
these *MEDICEAN STARS* I have found.   Cosimo II de Medici,
                                        r. 1609–21
And let me just add
that my pay's very bad –
are there any jobs going around?

A garden party

Did he foresee the future fight?
In this first book, he didn't write
a single word that made a case
for altering our central place.
And yet, the gauntlet had been thrown –
as Galileo must have known!
A sign that attitudes would harden
came in a friend's Bologna garden . . .   April 1610
This fellow had invited plenty
of influential *cognoscenti*

[117]

(if they were interested) to call
and see these things. They saw damn all,
and so of course these Bolognesi
decided that the man was crazy.
Kind Kepler sent an 8-page puff,
and asked him (reasonably enough)
if Galileo had some names
that he could use, to back his claims –
the Florentine, though, couldn't quote
a single personage of note
who would be willing to come clean
about the wonders he had seen!

Vague outlines in a coloured mist
would not convince the
   prejudiced.
It's to his credit that he saw
the things he is remembered for.

The Galilean Code

During the rest of 1610,
he sparked controversy again
by issuing more news – in code.
This was a method *à la mode*
for staking claims to something new
when there was more research to do –
to write it with the letters jumbled,
and hope that nobody had tumbled
before the finding was announced.
The first one couldn't be pronounced . . .
EUELOTPANPAREITEHRSTTTSERDHTIEOLVAMBSETITOV;

but when the jumble is undone,
it says that Saturn's 3 in I!
His second anagram was better,
and made a sentence (plus one letter) . . .
THESE PHENOMENAL VIEWS SHOOK US – D.

The letter makes it like a note
that some contemporary wrote!
(The 'D' could well have stood for Donne,*   *John Donne 1572–1631
who puzzled endlessly upon
the 'new philosophy', which brought
a new dimension to our thought.*   *And new philosophy calls all in
For if the world had started spinning,   doubt . . .
a revolution was beginning.)   'Tis all in pieces, all coherence gone
                                            The Anniversaries (1612)

The bombshell

What did these codes convey, when sorted?
The first one, as I've said, reported
that Saturn had appeared as triple.   In fact, he had half-glimpsed the ring,
But this was just a minor ripple   which puzzled him like anything.
compared with No. 2, which sent
a shock wave through the firmament!
Venus showed phases like the Moon.
This was a bombshell. Very soon,
the planet's ever-changing phase
was subject to the wondering gaze

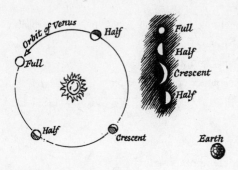

of others' eyes. What did it mean?
Well, two things . . . Venus must be seen
by sunlight it *reflects* to us —
it isn't just self-luminous.
And also, as the picture proves,
it is the *Sun* round which it moves!

Facing the facts . . .

This independent confirmation
boosted our hero's reputation.
The ones who'd scoffed at him, now saw
what was invisible before —
he went to Rome, where he was fêted
and generally adulated,   Summer 1611
and had a meeting with the Pope,   Paul V r. 1605–21
who asked about his telescope!
The doyens of celestial knowledge
in Rome's Jesuitical College

gave up what seemed a hopeless fight
to bolster up the Stagyrite.   Aristotle
The challenges were all too clear!
The Moon was *not* a perfect sphere,
but mountainous; four worlds rotated
round Jupiter, though he had stated
that we're the centre of rotation
(and always have been, since Creation);
and Venus goes around the Sun!
Had Galileo really won?

. . . but playing the game

He hadn't, as the whole world knows;
but till the starting whistle blows
there is no match . . . As I have said,
the *Messenger* was widely read –
but nowhere did he say straight out
that these new findings cast a doubt
on Aristotle's sacred views.
To be the first to spread the news
was his main aim, you can be sure –
and then, *res ipse loquitur!*   Let the facts speak for themselves
The point you need to grasp is this –
the Sun-centred hypothesis
had been a topic to discuss
since long before Copernicus –

but as a theory, nothing more.
That didn't go against the law;
but claiming to have *proof* would be
a very different cup of tea!
It's easy now, to say 'How blind!'
But you must always bear in mind
that Galileo couldn't *prove*
the solid Earth is on the move.
The crystal orbs beyond the Moon
would be discarded very soon –
but telescopic evidence
could still make admirable sense
with Tycho's plan. Perhaps the Dane

See p. 96

would come into his own again,
and see his 'halfway' concept fit
the sacred cows of Holy Writ!

Galileo writes a Letter

But Galileo's fierce polemic
made such refinements academic:
he got into a real mess
because of his Sun-centredness.
He wrote a Letter to Christina.
I doubt if he had ever seen her,
or if (to do the lady credit)
she took much notice, once she'd read it . . .

*Letter to the Grand Duchess
Christina, 1615*

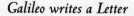

the contents were, in fact, addressed
to theologians and the rest,
to bring the business to a head.
'If there's a difference,' they read,
'between what rationalists know
and what the Bible says is so,
the ball is clearly in your court –
make Scripture chime with modern thought!'
The Cardinal to whom this went
was Robert Bellarmine, who sent   1542–1621
a firm reply . . . 'It's up to you
to prove Copernicus is true!
If you can do so, then we'll see
what updates there might need to be.'
A reasonable reply, in fact,
for proof was what our hero lacked . . .

A rap over the knuckles

But Galileo kept on nagging.
Deciding that he needed gagging,
some heavyweights (plus Bellarmine)
passed judgment on the Florentine.
Historians are still unsure
if he was ordered to *abjure*   Decree of 5 March 1616
the notion that we're not at rest,
or told it must not be *professed*.

One outcome of this fruitless fuss
was that poor old Copernicus,
now had his volume *banned*, instead
of simply mouldering unread.

The 'Dialogue'

Despite the toes on which he trod,
the latest Regent crowned by God
(Urban VIII) was so impressed
that Galileo was his guest
on six occasions: so you see
the March 1616 decree
by no means left him in the cold.
During these meetings, he was told:
'It would be excellent, I think,
if you took paper, quill and ink
and wrote about this new idea –
provided it does not appear
as if the case is watertight
to someone reading what you write.
As long as you don't ever say
that Scripture's blundered, write away!'
Armed with this papal dispensation
he wrote his book. Its publication
was subject to the Censor's view
on whether what it said was true;

Pope 1623–44

DIALOGUE ON THE
FLUX AND REFLUX OF
THE TIDES
Altered at the request
of
HIS HOLINESS THE
POPE
to A
DIALOGUE ON THE
GREAT
WORLD SYSTEMS
BY GALILEO GALILEI
FLORENCE 1632

but, being utterly perplexed
by Galileo's cunning text,
he gave it a cosmetic tweak,
and that was all. Three people speak . . .

Enter three characters in order of decreasing intelligence.

SALVIATI
I'm an Earth-moving modern, both learned and wise
[the author in fact, in transparent disguise].
I have to persuade these inferior two
that I'm right. In the end, they subscribe to my view!

SAGREDO
I stand for the man in the street, I suppose –
I know what the average citizen knows,
so I listen to what the protagonists say
and make the odd comment, to keep things in play.

SIMPLICIO

As you'll guess from my name, I am simple of mind.
I defend Aristotle and those of his kind
who believe in our Fixity. As you'd expect,
this ludicrous model is utterly wrecked!

SALVIATI

Simplicio put up a pretty good fight –
it required four days to convince them I'm right,
and the logic that really put paid to his case
was my Theory of Tides, not my findings in space!

SAGREDO

The theory is brilliant! Our artist has done
a plan of the Earth as it goes round the Sun.
Consider the speed of the surface at A
(i.e., midnight) and B (when it's got to midday) . . .
Though the rate of our spinning is constant,
    you'll see
that our orbital motion's *subtracted* at B
and *added* at A – if the great man is right,
we move slowly by day and then speed
    up at night.
Each increase in speed gives the oceans
    a push,
and the waters rise up with a general whoosh.
That's better than Kepler's ridiculous notion
that the pull of the Moon raises tides in the ocean!

Orbit around the Sun

B
Slower noon

A
Faster midnight

Kepler was right about the sea — he almost hit on Gravitee.

SIMPLICIO

Good friends, I concede – Aristotle's outdated!
This Earth-moving model that Old Nick created
explains the appearances better than mine.
Of course, we can't really probe matters Divine,
so the proofs you've been proving aren't *provably* true –
but I think that's for hair-splitting clerics, don't you?

SALVIATI & SAGREDO

We quite agree. While Rome pontificates,
let's feast ourselves. The gondola awaits.

[*Exeunt omnes.*

CURTAIN (for Galileo)

The Inquisition

Although the Pope had changed the title,
the printed dramatized recital
appalled him, when he looked inside.
They'd all been taken for a ride –
Simplicio, an utter twit,
was spokesperson for Holy Writ!
When Galileo came to trial,
he made a grovelling denial
that he had backed Copernicus.
'*Endorse* the guy? It's obvious
to anyone with any sense
that I reject his arguments!

He hasn't got a leg to stand on!
I'm telling people to abandon
this crazy Sun-centred idea!
But now that you have brought me here,
I must reluctantly admit
that having re-examined it,
my book does seem to under-stress
the *defects* of Sun-centredness –
perhaps, through trying to be fair,
I puffed it rather, here and there.'

'*Eppur si muove!*' ['*Yet it does move!*']

These *sotto voce* words of his
have echoed down the centuries,
though they were so much wasted breath.
They sentenced him to stay till death

There's no proof that he ever
  muttered
the words he's praised for
  having uttered.

within his villa's wall and gate:
he had about ten years to wait . . .
He stands for Reason, brought to heel
by Dogma; but I tend to feel
that Galileo should have known
he'd reap the whirlwind he had sown.
His prosecutors at the trial
let pass his fatuous denial,
perhaps because they weren't too keen
to martyrize the Florentine —
so, having brought things to the crunch,
he muffed his last defiant punch.
The oddest fact is this . . . Although
he wrote to Kepler, as you know,
and used his name to help the cause,
he turned his back on Kepler's Laws!
These did as much as he had done
to argue motion round the Sun,
because the new ellipses fitted;
but Galileo stayed committed
to all the epicyclic gear
that Kepler worked so hard to clear.
When he was rehabilitated,   In 1737
his mortal remnants were translated,
and now he shares our path through space
in Santa Croce's resting place.   Florence, 1737

[129]

(His book, the cause of this to-do,
was banned till 1822.)

Descartes and the Theory of Everything

The scene now shifts beyond the grip
of direct Papal censorship,
to cooler climates. Let us start
by mentioning René Descartes,
a French philosopher of note,
remembered for the well-known quote
I think, therefore I am. His dream
was a grand mechanistic scheme
that could explain why what we see
(from stars to ants) has come to be.
Even to contemplate this question
betrays a wholly new direction
in human thought – problems like these
were tackled by Xenophanes*
before the dreary interlude
of safe Dogmatic certitude!
Descartes dreamed up *x-y* notation
(graphs use this in their presentation),
and differential calculus,
which strikes us as miraculous,
considering he used to stay
tucked up in bed until midday.

1596–1650

He trusted reason more
   than sense,
although there had been
   precedents.
To Plato, things we think
   are real
are just a sketch of the Ideal.

*600 years before Christ's birth
he wrote: *All things derive from Earth,
the Sun and stars, the Moon and men –
and then return to Earth again.*

Like Balzac, he achieved renown
for working in a dressing gown.

A body's path, he thought, was *straight*,
unless obliged to deviate:
to keep the planets in their place,
he postulated whirling space –
not just a vacuum, of course,
but clouds of corpuscles, whose force
(through infinite collisions) serves
to push the planets into curves.
Which meant (to make the details fit)
the Cosmos must be infinite!

Seeking the centre

This genius had long been dead
when Newton, struck upon the head   Isaac Newton 1642–1727
by that green thought-provoking Apple,
decided that he'd have to grapple
with Gravity. The ancient view
(still widely held as being true)
was that an object's 'heaviness'
reflects its innate eagerness
to fall till it can fall no more –
in other words, it's reached our core,
the Cosmic centre, journey's end!
If we accept this innate trend,
a heavier stone has more desire,
so its velocity is higher

[131]

when matched against a lighter one.
(The famous Pisa drop was done
by someone who was bent on proving
that heavier things *are* faster-moving!)
As long as we were thought to lie
around the pivot of the sky
(the hub and focus of it all),
this model worked – a weight would fall
towards the Earth, the central place . . .
but if we're orbiting in space
(which now seemed an established fact)
there's much less in us to attract!

Newton's dilemma

And yet, attraction there must be . . .
That Apple falling from the tree
made Newton scratch his head and think.
If not sensate, what was the link
that made the Apple seek the ground?
Kepler, we've seen, had played around
with Gravity; we also know
that in the writings of Boulliau   Ismael Boulliau 1605–94
there's proof that he had contemplated
the Law that Newton formulated.

Newton's Law of Gravity

Two bodies (of mass M and m, let's say),
and separated by a distance d,
attract each other in the following way,
according to his Law of Gravity.

Their total mass (that's $M + m$, of course)
increases their joint power of attraction,
while d reduces the resultant force –
it's the denominator in the fraction . . .

$$Force \propto \frac{M+m}{d^2}$$

$$Force = \left(\frac{M+m}{d^2}\right) G$$

G is Gravitational Constant

. . . $(M + m)/d^2$. But it's better
to make things balance with an '=' sign –
to do this we require another letter,
the constant G. Now everything is fine . . .

Newton's disillusioned . . .

$(M + m)/d^2$ fitted
(well, more or less), but he admitted
that he found little satisfaction
in this idea of interaction
between two bodies far apart.
(That is one reason why Descartes
used corpuscles to get things going
and keep supplies of power flowing.)
But Newton, stuck with his equation,
could not resort to such evasion,
and got increasingly annoyed,
faced with a planetary void
and no known method of transmission.
This unsustainable position
would later wring this heartfelt plea:
'Do not pin Gravity on me!'
It didn't work, in any case –
the Moon refused to keep its place . . .

Third Letter to Richard Bentley,
11 February 1693

All at sea . . .

At last the Moon obeyed his law,
which it had never done before;
but that's because, two decades later,
he got his hands on better data.

Newton had abandoned work
on gravity *c.* 1665

Sir Christopher Wren — 1632–1723
went to dine with some men,
and he told Drs Halley and Hooke — Edmond Halley 1656–1742
that the first one to prove — Robert Hooke 1635–1703
how the planets must move
would receive a magnificent Book!

Robert Hooke said: 'I *know*
how the wretched things go –
it's an inverse-square law they obey
under Gravity's pull.
You shall have it in full . . .
What's the volume about, by the way?'

But the book was unclaimed
by the deadline Wren named,
although Halley still puzzled. He thought
that he might get a clue
from a Fellow he knew
who had rooms over Trinity Court . . .

So to Cambridge he went.
Newton told him: 'I spent
a lot of time fooling around
with this Gravity stuff,
till I'd had quite enough!'
But his scribblings couldn't be found . . .

August 1684

And that was the start
of the prominent part
Edmond Halley would play, for he made
this most awkward of men
work it all out again,
took the book to the printer – and paid!

The reason was tied up with Trade . . .
Bold ocean-crossing trips were made,
but using sand-glasses as clocks
sent many vessels on the rocks –
a mere five minutes fast or slow
changed by 300 miles or so
the ship's location east or west.
This tricky problem was addressed

as follows . . . Every thirty days
the Moon completes its change of phase
by orbiting the starry sphere –
as in the picture printed here!
Every two hours, as you see,
it travels east by one degree.
And so, aboard his tossing boat,
the navigator made a note
of just how far it was away
from some star (Spica, let us say),
then, using tables, calculated
when they must have been separated
by this amount. From this, he'd know
the right time several hours ago.
The calculations covered pages,
involving several separate stages –
made harder, as you would expect,
by wondering if they'd be wrecked
before he'd finished them, and learnt
exactly where they were (or weren't).

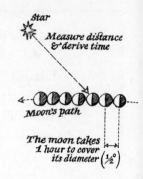

α Virginis, a so-called 'Clock Star'

A prime site

The English and the French were vying
in this new business of supplying
good lunar data in advance.
The new Observatoire de France

A Frenchman, Monsieur
    St Pierre,
came up with the *Méthode Lunaire*.
King Charles's mistress was
    recruited
to urge it on him when it suited.

(the first creation of its kind) Completed 1672
meant England had been left behind,
so Charles II there and then r. 1660–85
entrusted everything to Wren,
who found a site in Greenwich Park,
well out of town, and nice and dark.
Put up in haste, it was a bodge, In use by June 1676
built on a royal hunting lodge
(it should have faced north-south, east-west,
but doesn't, though Wren did his best).
However, from this princely site
the heavens were measured, day and night,
and Halley gave its boss* no peace *John Flamsteed, first
till he consented to release Astronomer Royal, 1646–1719
the data Newton thirsted for
if he was to confirm his Law.
The Moon turned out a perfect fit —
so he and Halley had a hit! *Principia* published 1687

Free Fall

Why is the Moon's path curved in space?
What keeps it firmly in its place?
These questions Newton had to face!

A body keeps its straight-line course
unless deflected by a force . . .
Our Gravity *must* be the source!

Ascend an elevated site
and throw a stone with all your might
in absolutely level flight . . .

The Earth's pull *g* comes into play.
The stone starts falling straight away
and lands upon the ground at *A* . . .

Now use a catapult, and see
how far it goes . . . It lands at *B!*
If it's launched fast enough, maybe

(our planet being like a ball)
the curving ground will match its fall –
and it will never land at all!

This was Sir Isaac's way of showing
why circling bodies keep on going.
But God must organize the throwing . . .

Invisible, unknowable, indispensable . . .

One question answered raises two.
It follows (if this statement's true)
that every new discovery
increases our uncertainty!
This seems to have been Newton's case . . .
A force that pulls through empty space
defied belief. Some medium,
some Spirit in the vacuum —
above our minds, beyond our sight,
must transmit Gravity and Light!
This 'ether', though outside our sense,
became a frame of reference
to which all movement was referred.
The concept now appears absurd,
since Albert Einstein put the boot
in motion that is absolute;
but nobody was then prepared
for E to equal mc^2 —
and Newton only starts to fail
when used upon a Cosmic scale!
Another snag he raised with Bentley See *p.* 134
was this . . . Beginning very gently,
the stars' own mutual attraction
should start a general contraction
(inversely as d^2) and cause

a massive crash! Had his own laws
been overruled in stellar space?
Did every star remain in place
within its chosen constellation
through God's especial dispensation?
The question that was asked much later
(should Science bow to the Creator
and let Him have the final say?)
is still in people's thoughts today!

The stars are all in orbit too —
a fact that Newton never knew.

7 Measuring
(1671–1838)

After the medieval night,
with Superstition put to flight,
a light too brilliant to set eyes on
bedazzled Reason's far horizon . . .
Set free from our sublunar sphere,
we had the run of space; and here
there is no frontier at all –
the pull that made the Apple fall
is just as Cosmic as the force
that keeps the Earth itself on course.
All's one in the Newtonian plan –
the same laws rule Sun, Moon and Man!
So measure, measure . . . that's the key
to rational Cosmology!

1 THE DISTANCE OF THE SUN

π *in the sky*

To find this fundamental distance*
involved phenomenal persistence:
some Seekers after Truth would die
in their attempt to measure π!
You may well ask: Why such a fuss
about our orbit's radius?
The reason's this. Kepler's Third Law
(which I have talked about before,
so check the facts if you're unclear)
links *P*, a planet's length of year
to *a*, its distance from the Sun
(calling the Earth's own values 1).
But from his Law we merely know
their distance as a ratio
of our own distance: that's to say,
we need to know *our* planet's *a*
in leagues or miles or whatever.
And so began the great endeavour . . .

*The Astronomical Unit

The solar parallax, i.e.,
just half the width the Earth
   would be
seen from the Sun. A beer-mat,
   say,
about a kilometre away.
See *p.* 000

Measuring the Earth (the first job)

A French *savant* named Jean Picard
determined how far round we are —
a length they simply *had* to know
before they started — see below!

1620–82

Measuring the Earth 1671

Like Al-Farghani, Picard chose
two points, from where his guide stars rose
(or fell) by one degree, when viewed
along a line of longitude.
The arc he based his measures on
began just south of Amiens
and passed through Paris, on the way
to end near La Ferté-Alais.
He gauged its length by taking sights
through primitive theodolites,
and got a length for 1°
(essential for Cosmology)
closer than 0.1%
to the accepted measurement!

See p. 55

Picard's radius 6372 km;
modern mean radius 6371 km

Trying the Mars method (1672)

The human record of stargazing
contains few ventures more amazing

than Jean Richer's intrepid mission —   1630–96
a transatlantic expedition
to South America (Cayenne),
returning safely home again —
not something one would lightly do
way back in 1672.
September of that fruitful year
saw Mars exceptionally near;
Richer was hoping to detect
a tiny parallax effect
when he compared his measures later
with those of his collaborator
J. D. Cassini, who remained   Jean Dominique Cassini 1625–1712
in France. The value they obtained
(the first to be derived for π)
was only 8% too high!   9.5" instead of 8.8"

Measuring with Mars

Cassini (in Paris) and Richer (Cayenne),
at a distance denoted by D,
gauged the angles that Mars
made with various stars
to 1/500th of a degree!

When Richer got back, their results were compared.
They must have been tickled to bits!
Richer's expedition
got a *different* position
(Let me check that the working-out fits) . . .

If D was 1,800 French leagues*
(which can't be too wide of the mark),
and Mars moved by μ
when comparing the two
(let's say 25" arc) . . .

Then its distance, called S, is $D/\sin \mu$† –
15 million French leagues (which I've used
because leagues had their say
until Charlotte Corday‡;
so I hope you don't get too confused).

From Kepler's Third Law we know how to relate
the distance of Mars from the Sun,
with the Earth's. As I've shown,
once the *difference* is known,
then the battle is more or less won . . .

Call our distance a_e, Mars's distance a_m . . .
then $a_m = a_e + S$.
So a quick substitution,
and there's the solution –
35 million leagues, more or less!**

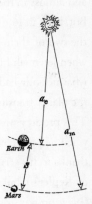

*7000 km
†Strictly speaking ½ $D/\sin$½ μ
‡1768–93

**About 140 million km.
The true value is
149.6 million

When Venus comes between us ...

Other astronomers would try
to see if they could nail π
by planning similar attacks
upon the solar parallax.
But their results did not agree,
nor did they lend validity
to Richer & Cassini's figure.

From 70 million–200 million km

If only μ (the shift) was bigger
(i.e., the distance S was less),
they'd have more chances of success!
Venus comes closer, it is true,
but then she's almost lost from view
because the Sun's extremely bright
and almost in the line of sight.

But, back in 1639,
the two of them were bang in line,
when Jeremiah Horrox saw
what nobody had seen before –
the planet transiting the Sun.

c. 1618–41

This prodigy was 21
when he predicted the alignment,
thanks to his scrupulous refinement
of Kepler's tabulated data

The *Rudolfine Tables*

(Newton was grateful to him later,

when he was struggling with the Moon).
But Death waylaid him all too soon –
his tragic end at 23
impoverished Cosmology.

As Newton gratefully confessed,
his lunar tables were the best.

Fights for sites

These transits are extremely rare.
Two, spaced by eight years, form a pair –
and then a century to wait!
Halley sat down to calculate

His paper published 1716

when we would see another two –
the first one of the pair was due
in 1761, the second
in 1769. He reckoned
(knowing full well that he'd be dead)
that if observers could be spread
from pole to pole, then the *duration*
would change as seen from every station,
because the planet's parallax
would give it slightly different tracks.
Britain and France sent expeditions
to distant sites in prime positions –
in '61 they were at war,

Seven Years War, 1756–63

but Venus was worth fighting for . . .
one British vessel was attacked
(11 people killed, in fact)

when it had barely put to sea.
Brits, too, behaved disgracefully
towards some harmless folk from France
who, by a most unlucky chance,
were spotted on their island site;

Rodrigue, Indian Ocean

unable to put up a fight,
they saw one boat go up in flames,
and lost the other, so their aims
were less concerned with finding π
than looking out for passers-by.

Frustration, privation, colonization

A bolder pen than mine should trace
the drama of this epic chase!
Guillaume Le Gentil missed his date

1725–92

in '61 – the ship was late,
which left him on the sunny ocean,
defeated by the vessel's motion,
watching the planet's silhouette
like some delectable coquette.
To make quite sure, he stayed out east
for eight years, but the cloud increased
when, on the second Day of Days,
she veiled her outlines from his gaze!
Jean Chappe, who braved Siberian snow

1722–69

(since that was where he had to go

to see the '61 event),
performed so well that he was sent
to watch the second of the pair
from distant Mexico; and there,
as fever gripped and life receded,
he made the observations needed
to register the planet's track.
One member of his team got back . . .
Another group that went to look
were chaperoned by Captain Cook
(although he had, in fact, been sent
to seek a Southern Continent).

1728–79

Australia sighted 20 April 1770

π IN THE SKY

The Ballad of the Transit Chasers, 1761

When Venus comes between us,
then π is in the sky!
Wake up, captain, if you're sleeping down below!
Take us south of the equator
to collect precision data,
and escort us homeward later
(that is, if the Creator
wants to bring us back from where we've got to go!).

[151]

If we're to measure Venus
and derive elusive π,
then we need to know exactly where we've gone!
We know that we are *here*
(that much is fairly clear);
but with no known landmarks near
we've but the haziest idea
of the true position we are standing on!

The distance D between us
and the other slaves of π,
cannot be found without our *longitude*.
If we had GMT
what a doddle that would be;
but the 18th century
lacks precise chronometry –
so our methods may appear a trifle crude!

Before she comes between us
and we have a stab at π,
our timekeeping has got to be corrected.
We've brought a clock that goes,
but no one really knows
if the hour that it shows
is the time that we suppose –
so a small observatory must be erected!

Therefore, while we wait for Venus
and prepare to measure π,
there's a list of pre-conditions to be met.
Phew – that's all the preparation!
Thanks to smooth co-ordination
we've established our location,
and we're full of expectation
as we strain to see her longed-for silhouette!

Now Venus *is* between us!
We are going to pinpoint π
by noting when she's fully on the Sun . . .
She is moving very fast . . .
Has the moment come at last?
I'm not sure . . . maybe it's passed?
Yes, I've missed the timing! *Blast!*
Still, in eight years' time we'll see another one!

π defiant . . .

As our reflective balladeer
has made dramatically clear,
the transit method lacks precision.
A mere five seconds' indecision
when Venus has at last begun
to draw her chord across the Sun
(with equal doubt about her exit
some hours later) simply wrecks it!
As well as dodgy observations,
none of these enterprising stations
used clocks that had been synchronized;
so you will hardly be surprised
that puckish π remained in doubt.
It wasn't really sorted out
till David Gill (as I shall tell)
camped at Mars Bay with Isobel.

2 LOOKING UP

How far to the stars?

Splitting peas (again)

When London was engulfed in flame   1666
(a baker's oven was to blame)
one burning question of the hour
was this: given sufficient power,
might an astronomer detect
a six-month parallax effect

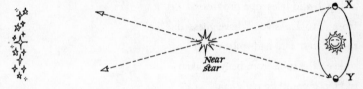

when stars are viewed from *X* and *Y*?
Tycho, we know, had had a try   See p. 97
and failed; so it had to be
much smaller than a garden pea –
but now, with optical assistance,
it might be spotted, with persistence!
Hooke, one of those who nursed this hope,
designed a *zenith telescope* . . .
A telescope secured tightly
will see the same stars passing nightly,
for it is anchored to the ground
and turns round as the Earth turns round

(they pointed upright, since the air
degrades the image least up there).
A target star was then selected;
two spider-threads that intersected
were used to check its nightly tracks.
A drifting caused by parallax
would make it miss the intersection,
so they repeated their inspection
whenever circumstance allowed
(i.e., there wasn't too much cloud)
as Earth and telescope progressed
from X to Y. As you'll have guessed,
the greatest care had to be taken
to keep the tube from being shaken –
in fact, it could be truly said,
success hung on a spider's thread.

Wren (again)

Hooke led the parallax attack
by fixing, to his chimney stack,
a tube that passed through roof and floor
(ten metres long, or maybe more)
in his own rooms at Gresham College.
As he was later to acknowledge,
the building wasn't really stable,
and therefore he was quite unable

In 1669

to work out if the star* had drifted,
or if the chimney stack had shifted!
His colleague Wren was also keen
to check what movement could be seen –
the Monument, which he designed
to keep the famous Fire in mind,
not only celebrates the Blaze
but is tribute to the craze
for looking overhead at night
to prove Copernicus was right,
for there's a basement you can enter
and look straight upwards through its centre!
Wren also saw a chance to use
half-built St Paul's for zenith views –
the south-west tower's roofless shell
contained a lofty staircase well
(designed to house a monstrous clock)
which was as steady as a rock.
But nothing ever came of it –
the lens he'd planned on didn't fit.

*γ Draconis

Completed 1679

Remember, nobody had *proved*
it was the solid Earth that moved.
That's why they kept up their attacks
upon the stellar parallax.

1675–1711

Too far, too fast, wrong way . . .

A good half-century was spent
in futile experiment.
Observers claimed to have detected
the sort of movement they expected

for stars they thought were near to us
(such as the brightest, Sirius);
however, instrumental flaws
and human error were the cause.
And then, beside the Thames at Kew,
the house of Samuel Molyneux   1689–1728
supplied the firm foundations needed.
James Bradley, Samuel's friend, succeeded   1693–1762
in tracking Hooke's draconic star . . .   1726
The trouble was, it moved too far,
too fast, and in the wrong direction!
Despite the most minute inspection,
Bradley found nothing to suggest
his telescope was not at rest;
the star *was* shifting in the sky,
though maybe God alone knew why!

Seeing the light

The answer came to him one day,
but in an unexpected way.
The fact that light has finite speed
(although it's very fast indeed)
was noticed by a clever Dane,
Olaus Römer (I explain   1644–1710
the reason in the *Light's Flight* note).
Well, Bradley went out in a boat,

and registered the curious fact
that every time the vessel tacked,
the masthead pennant and burgees
altered their angle to the breeze.
He wondered how that could be right . . .
then, in a blinding flash, saw Light!
The way in which the pennant blew
was caused by (1) the wind, and (2)
the vessel's own velocity
(denoted here by *A* and *B*).

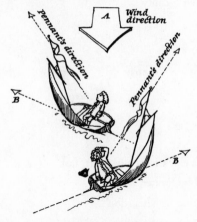

Therefore its angle wasn't showing
the point from which the wind was blowing,
although it *seemed* to be. Of course!
The starlight sent out by the source
was shifted from its true location
by what he called the 'aberration'

Light's flight

Jupiter's shadow is immense:
its moons just vanish, it's so dense.
Before the days of Greenwich pips,
the start or end of each eclipse
was calculated in advance –
clock-watchers therefore had a chance
of checking what the time must be,
whatever their locality.

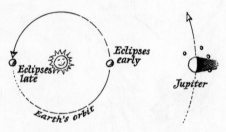

But Römer's clock went slow, then fast,
when timed by Jupiter. At last
he realized that its 'alteration'
followed the changing separation
between the satellite and us –
light wasn't instantaneous . . .
More distance meant more time to wait,
which made all the predictions late!

caused by our motion. Even though
(compared with light) we're very slow,
the shifting that he had detected
matched how it ought to be deflected,
and proved our motion round the Sun –
delighting almost everyone.*
Alas, things did not go so well
for his co-worker, Samuel,
who died of some obscure condition –
his widow left with the Physician.

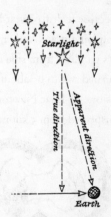

*That's when Manfredi wrote a book
claiming that he (as well as Hooke)
had spotted parallactic shift.
The Censor was extremely miffed –
the Roman Church still disapproved
of anyone who said we moved.
(Eustachio Manfredi 1674–1739)

The apple plucked . . .

More than a century went past
before a star was plumbed at last
(the brilliant feat achieved by Bradley
affected parallaxers badly –
the aberrational effect
made true shifts harder to detect).
In fact, the world would have to wait
till Bessel (1838)
tackled a faint star in the Swan.
This object was decided on

Friedrich Bessel 1784–1846

61 Cygni

because it drifted year by year,

'Proper motion'

implying it was fairly near.
Its parallax, from side to side,
turned out to be about as wide
as one of Isaac Newton's eaters
inspected from 3 kilometres —

The same size as the one he's gnawing
in David's splendid cover drawing.

600,000 times as far
as we are from our sunny star!
But I have leaped out of my text —
it's Herschel I must turn to next.

8 Heaps of Data – Answers Later
(1770–1900)

Blow this . . .

Forced marches, heat, and rotten pay
had been the order of the day
for Hanoverian musicians
who'd passed the onerous auditions
to get into the band, and puff
their mates to victory. 'Enough!'   Seven Years War (again)
thought Wilhelm Herschel, oboist,   1738–1822
'I was a *Dummkopf** to enlist –   *Fool
the army wasn't meant for me,
or I for it. I'll cross the sea
to England!' So he upped and went,
abandoning his regiment
after the Hastenbeck disaster,   26 July 1757
and ended up as organ-master
in fashionable Bath. He taught,
did composition of a sort

(he had some unsuccessful goes
at writing oratorios);
and then a volume* caught his eye
about the wonders of the sky . . .

*Ferguson's *Astronomy*

Per DIY ad astra . . .

His new ambition quickly gelled –
to see what *no one* had beheld!
But bringing fainter things in sight
required large amounts of Light
to make them stimulate his vision,
prompting the radical decision
to *make* the telescopes he needed.
So in his outhouse he proceeded
to build a forge, run it full blast,
melt copper mixed with tin, and cast
metallic mirrors of such size
they broke the bulwarks of the skies
(or so the chiselled words would say).

Coelum perrupit claustra

The largest telescope today
still used in Britain, could be put
inside his famous 'forty-foot',
along whose tube the Monarch led
his topmost cleric, having said:
'Come, my Lord Bishop,* follow me –
I'll lead you to Infinity!'

George III r. 1760–1820

*The Archbishop, complete
with mitre,
might well have found the
going tighter.

Now he was famed; but what began it
was noticing the seventh planet
in 1781. He took
a leaf from Galileo's book
and named it after George III   Georgium Sidus
(though 'Uranus' would be preferred).
This flattery went down so well
(he was no *Dummkopf*, you can tell)
that George, to thank him for the thought,
made him Stargazer to the Court,
which led to tedious requests
to show the stars to royal guests.

The end is nigh ...

He swept the firmament all night!
To keep his dark-adapted sight,
his sister Caroline would sit   1750–1848
beside a lantern she had lit,

and note down what he saw and when,
although the ink froze on her pen.
Pursuing his celestial trawl
he made an unexampled haul . . .
clusters of stars and hazy patches
he published in colossal batches –
a thousand at a time! His dream
was to decode the heavenly scheme . . .
to analyse their distribution
and study stellar evolution.
He sampled densities (e.g.,
how many stars per square degree),
observing great unevenness.
Had they begun to coalesce?
Had interstellar gravitation
begun this vast coagulation?
The break-up of the Milky Way
was bound to come about one day
if star-groups started to condense.
The implications were immense!
He wrote: 'The fact that it's decaying
is just another way of saying
that something must have *started* it –
its lifetime isn't infinite!'
Newton, foreseeing such collapse,
thought that some agent (God, perhaps)

would come up with a counter-force
to halt its fatal inward course;   See p. 141
and even Einstein was compelled
to stoop to λ, which repelled!   See p. 206
But Herschel did not seem averse
to winding down the Universe.

Any ideas?

Stars were, he thought, of different ages,
commensurate with various stages
from infancy, through youth, to prime —
a new conception at that time.
(Among the objects he could see
were misty spots called 'nebulae';
he thought each evanescent glow
to be a star in embryo.)
But this Columbus of the skies,
although he tried to theorize,
had far outstripped the knowledge base
on which to build a solid case.
At that time no one had a clue
how stars shine; no one even knew
that other galaxies exist;
the pulpits, to a man, dismissed
the evidence hacked from the rocks
that presaged major seismic shocks

Most of these nebulosities
are now known to be galaxies;
but he was right in his deduction
that nebulae cause star-production.

when fossils proved that they *pre-dated*
the day our planet was created!
We've reached the time when Observation
eludes the grasp of Explanation –
it won't be till the First World War
that Theory catches up once more . . .

Dramatic, chromatic . . . prismatic!

With William's death, his offspring John   1792–1871
took up the reins and carried on,
taking his wife and growing brood
to Cape Town's distant latitude,
where finds beyond his father's eye
were netted in the southern sky.   1833–8
But now it's time to meet the Prism,
which caused as great a cataclysm
as Galileo's spyglass did!
This pleasing shape wrenched off the lid
that hid the way the Cosmos ticks –
without its technicolour tricks

we wouldn't have much more idea
of why stars form (and disappear)
than Herschel did; nor would we know
about the Big Bang long ago!
Its influence was so dramatic
I've done a box on things Prismatic . . .

Inside the rainbow

If daylight's made to pass
through a triangle of glass
you will see a band of colours (blue to red),
which, to our human sight,
gives a single colour, White.
But if you use a *spectroscope* instead . . .

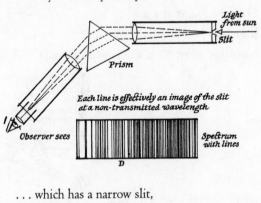

... which has a narrow slit,
and a lens to focus it,

and you grab a ray of sunlight (when it shines),
the band of light you'll see,
if it's set up properly,
will be crossed by countless very narrow lines.

'Like a barcode!' you might say,
which is quite true in a way,
since each element forms lines that are unique.
You can find them one by one
in the spectrum of the Sun,
provided that you know which lines to seek!

But Prisms are outmoded;
the light is now decoded
by a plate like a CD, with rainbow tints.*
Flash the gadget at a star . . .
learn how hot, how fast, how far,
how large, when born, and how it's altered since!

Dist: 250 light years
Temp: 6500°
Size: 2000000 km
Speed: 25 km/sec
bogof offer?

*They're ruled with unseen lines, creating
 what's known as a *diffraction grating*.

Reading the lines

Kirchoff, in 1859,   Gustav Kirchoff 1824–87
got interested in a line
crossing the spectrum of the Sun.
It was a most conspicuous one,
distinguished by the letter *D*.
Kirchoff proved its identity
to be unquestionably the same
as that seen in the yellow flame
when common salt is set alight
(although, this time, the line is bright).
He knew the line to be produced
by *sodium*, so he deduced
(after a lot of careful work)
that sodium must also lurk
within the Sun! Perhaps he'd find
what other elements had been assigned
to sunlight, if he kept on going,
and checked their spectra when they're glowing?
His colleague Bunsen's burner blazed   Robert Bunsen 1811–99
as temperatures were duly raised,
and samples fizzled as they sent
their barcode through the instrument . . .
Iron gave some conclusive hits;
he also found convincing fits

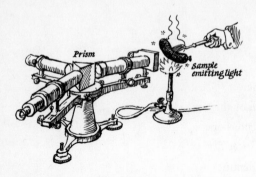

for nickel, and magnesium too,
and copper, which shines greenish-blue;
chromium, calcium and zinc
turned up as well; and who would think
that *barium* would be around?
(This makes all nine that Gustav found.)

Enter the Amateur . . .

His brilliant research paved the way
for what astronomers today
call *astrophysics*. Spectra show
what otherwise we could not know
about a star (how large, how hot,
its evolutionary stage – the lot);
added to that, they let us see
the secrets of the nebulae,
and light that's older than our planet
tells of the star-swarms that began it!

But this is to anticipate . . .
At that preparatory date,
few governments would underwrite
research so pure and recondite —
the institutions they'd created
were principally dedicated
to measuring the Earth's rotation
(the key to time determination),
and they were definitely sticking
to keeping clocks and watches ticking!
So early Astrophysicists
(and others) wooed philanthropists
if they did not possess enough
to buy the necessary stuff.
The time may never come again
when *scientific gentlemen*
(and ladies too, as we shall see)
do cutting-edge Astronomy.

William finds a vocation

From such a range of names to sample,
let me select a prime example . . .
When well-off William Huggins read   1824–1910
what Gustav had achieved, he said:
'It was as though a desert flower
had bloomed after a sudden shower!

A host of possibilities
was let loose by this work of his!'
He was among the very first
to get the light from *stars* dispersed –   In 1862
and in their feeble radiation
he saw an ordered graduation.
White stars had lines that were much broader
than in the Sun (the solar order
was matched by stars of *yellowish* hue);
red stars, like white, had wide lines too.
To see the spectra that they showed,
he heated gases till they glowed

with voltages of lethal force
(no Safety issues then, of course).
The white stars' pattern indicated
that *hydrogen* predominated;
the lines of other elements
were also there, but less intense
than in stars of the yellow sort.
Here, as in Gustav's first report,
the lines of *metals* showed up clearly —
just like the Sun, or very nearly.
The red stars' patterns weren't so clear,
but *compounds* started to appear —
two elements, or even three,
in intimate proximity.
So red stars are the coolest kind,
since compounds cannot stay combined
if they're too hot — a vital hint
that temperature is linked to tint.

The H-R Diagram will show
what colour tells us — see below!
(*p.* 191)

Beginning spinning . . .

A scientific Jesuit,
Angelo Secchi, did his bit   1818–78
in this immense ungarnered field.
Five years of patient work would yield   1863–7
four thousand careful observations
of likenesses and variations

[175]

among these ghostly stellar strips.
To sort out their relationships
he graded them by *spectral class*.
This raised the question: Do stars pass
through evolutionary stages?
Do different tints mean different ages?
If they do, are the youngest white —
or are they red? Was Herschel right
to say that stars start off as gas?
The famous theory of Laplace
(his Nebular Hypothesis)
suggested such a genesis . . .
The Sun, he thought, had its beginning
inside a gas cloud, which was spinning.
It threw off rings as it spun quicker,
the central part began to flicker
and turned into the shining Sun —
the rings, meanwhile, had begun
to form short arcs which coalesced . . .
the planets, as you've doubtless guessed!
If Angelo had been correct
to say that many stars he checked
have Sun-like spectra, this implied
a host of planets Cosmos-wide.
We are by no means an exception —
Life's part of the Divine Conception!

Pierre-Simon Laplace 1749–1827

1796

Progressives took the fact
for granted
that Mars and Venus were
implanted,
and ardent Pluralists insisted
that even Jovians existed.

Solar heating

Such theories, though, were not much good
until their authors understood
the basic problem — what's the key
to stars' supplies of energy?
In this debate, a major player
was Dr Julius von Mayer.

1814–78

A Sun made out of coal, he thought,
would find its fuel running short
after five thousand years or so,
which isn't long enough ago;

Since there's no oxygen in space,
it wouldn't burn in any case

so, having recognized its flaws,
von Mayer dreamed up *meteors*
that strike the Sun with such a crash
they give out a terrific flash.

Meteoric theory 1848

Don't think von Mayer was
    a fool . . .
He showed, at the same time
    as Joule,
that heat can be derived from
    motion —
but Joule got credit for the
    notion.
(James Prescott Joule 1818–89)

To keep it nice and hot, he reckoned
that 90 billion tonnes per second
must go this way; but if that's so,
the things would also fall like snow
upon the Earth — we would be pelted
until our solid crust had melted!
Hermann von Helmholtz started thinking,   1821–94
and wondered if the Sun was *shrinking*?
A gravitational collapse
which made it smaller by perhaps
a hundred metres every year
(invisible, of course, from here)
meant energy would be released
for 20 million years at least   Contraction theory 1854
after it reached its full-blown state —
although they hadn't long to wait
before it had completely shrunk,
and when that happened, they were sunk.
But fossils found by Mary Anning   1799–1847
implied a solar output spanning

a *hundred million* years or more
(which Darwin multiplied by 4   Charles Darwin 1809–82
to make sure that we had evolved)!
The problem would remain unsolved
until E equalled mc^2,
for which, I hope, you are prepared . . .

π *in the sky (again)*

A century had passed away
since those attempts to measure a
(the Earth-Sun distance, you'll recall),   See p. 148
when π's pursuers did their all
to measure Venus as she passed
across the Sun. And now, at last,
repeat performances were due
in '74 and '82.
Provisioned to survive a year,
and loaded to the line with gear,
her votaries set sail once more!
In all, in 1874,
some eighty hopeful groups were sent
to get the most from the event.
Six each from France and Germany;
twelve British, three from Italy;
eight from the USA, one Dutch;
but none of these amount to much

[179]

compared with Russia's 26!
(It wasn't π but Politics
that made their governments decide
they simply couldn't stand aside
and let the others take the glory.)
Alas, it was the same old story!
Observers got the timing wrong
because they paused a touch too long,
or in some cases jumped the gun;
and so the distance to the Sun
was not in any way refined.
These ventures also underlined
the need for using common sense
in undisturbed environments
such as Kerguelen, which is not
a very touristy spot,
surrounded by the Southern Ocean . . .
The British had the clever notion
of taking stewing rabbits there –
they've stripped this untouched Eden* bare.

*Used in the sense of 'undefiled'.
Kerguelen's windswept, wet
and wild.

GILL'S EXPEDITION TO ASCENSION, 1877

MARS WHA HAE . . .

The Ballad of David and Isobel

All that money down the drain!
Venus made us fools again!
Transit methods are insane,
as this farce has shown!

Makes me positively ill,
knowing π defies us still.
Say one day to Mrs Gill:
'We'll do it on our own!

'This September, Mars will be
closer than in history . . .
Isobel, you'll go wi' me.
We'll fix this milestone!'

Work like hell to raise the dough.
Need £500 or so . . .
Ascension is the place to go –
π's cover will be blown!

Measure Mars at dusk each night,
then again, before it's light.
(In the picture to the right,
the *rationale* is shown.)

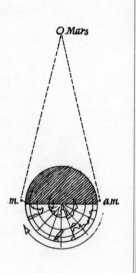

Twenty tons of gear I'll need
if my method's to succeed.
Anchor up, and may God speed!
Off to the unknown!

So far, clouds are all I've seen!
Weather's worse than Aberdeen!
Wife tells me: 'Tonight I mean
to leave you all alone . . .

'. . . and walk directly south from here.
Perhaps the stars will reappear!
I'll take a man (or two), don't fear,
to act as chaperone!'

As the Sun begins to rise,
back she comes, with shining eyes.

'David, such pellucid skies –
and how the night has flown!'

Pack the gear up straight away,
cart it to her cloudless bay,   Now Mars Bay
re-erect it . . . *Mars wha hae!*
Soon the planet's shown . . .

. . . parallactic oscillation!
What a perfect situation!
Camping in such sweet privation,
out here on our own!

Here and there, along the track
to the telescope and back
(moonless nights are very black)
she lays a guiding stone.

Earth is leaving Mars behind . . .
Nightly shift is much declined . . .
Sail home, to thresh and grind
the harvest we have sown!

Scots wha hae – we've humbled π!
Transits made it much too high!   Mostly 8.81"–8.88"
After all that hue and cry,
one pair of eyes alone . . .

... belonging to a frugal Scot
achieved what fleets of ships could not!
£500 paid for the lot –
our parallax is known!

Gill's figure 8.78"

Modern figure 8.79"

That phantom stuff ...

The curious way that light behaves
suggests innumerable waves
rippling outwards from the source.
But waves cannot pursue their course
without something to ripple through,
and so the *ether* concept grew –
a medium for propagation
of interstellar radiation.
It gives no clue to its existence ...
No mass, no colour, no resistance,
and motionless by definition –
the *Universe* can't change position!
Cleveland, Ohio, USA
is where two people saw a way
they might unzip this phantom stuff,
if their technique was good enough.
But Michelson and Morley failed –
and *Relativity* prevailed!

Albert Michelson 1852–1931
Edward Morley 1838–1923

1887

Opening the can

The ether, though beyond our sense,
was taken as the reference
to which all motion was referred.
The flight of a migrating bird,
the Earth's ellipse around our star . . .
however fast, however far,
the place the object was located
could readily be calculated.
Euclidean geometry
was all you needed – Q.E.D.!
Space was a kind of 3-D grid,
till M. & M. prised off the lid,
releasing lots of *lumbricina*   Worms
and opening a new arena
where astrophysicists would seize
innumerable Ph.D.s . . .

$c - S = c$ (?)

Using our planet as a base
(it's moving pretty fast through space),
and working on a marble block,
which was as steady as a rock,
they kept an eye on light's speed (c)
along the Earth's trajectory

(call our speed S^*). A beam of light
aligned with our ethereal flight,
would travel *slower*, they expected,
compared with one that's been reflected
at right angles. That's because S
makes light's speed seem a fraction less
when in configuration A,
since both are going the same way
and therefore S reduces c;
but in configuration B,
when S has been reduced to Nought,
light would seem faster. So they thought!
Soon they were asking: '*What the hell?*'
As far as M. & M. could tell,
light always went at c, full stop!
They'd got it wrong — it was a flop!

*About 30 km/sec

The speed of light *in vacuo*,
as I am sure my readers know.

Unthinkable — but shrinkable

This bombshell was as great a blow
to the unquestioned status quo
as Kepler's shocking revelation
that circles were without foundation.
It was impossible — but true!
Not knowing what they ought to do,
M. left it there, and so did M.;
but two researchers followed them,

One M. would get the Nobel Prize
for other projects he'd devise.

namely FitzGerald and Lorentz,
who managed to make perfect sense
of this disturbing paradox . . .
Suppose that the observer's clocks,
when used to time the light, went *slower*.
Then, if its speed was really lower,
it wouldn't show, since it was reckoned
in terms of an *extended* second,
which compensated perfectly
and kept *c* what it ought to be.
They also said that fast things *shrink*;
but that is quite enough, I think,
to show that we have left behind
the intuitions of the mind
conditioned by the world we know,
where things are Relatively slow!

George FitzGerald 1851–1901
Hendrik Lorentz 1853–1928

This crude attempt to give the gist
will horrify a Physicist.

Scan the sky with AgI . . .

The Prism was a mighty stride . . .
but so was silver iodide,
which turns black when exposed to light —
so bright is dark, and dark is bright!
In 1839, Daguerre
explained in full how to prepare
a silver plate that's squeaky-clean,
and hold it over iodine

Louis Daguerre's daguerreotype
process. (1787–1851)

to sensitize it with its vapour.
A method giving prints on paper
(by Archer, 1851)
became the universal one
for twenty years, although the plate
would lose its light-receptive state
once dry: this snag explains the need
to coat it, and at once proceed
to take the picture soaking wet.
This made life difficult; and yet
those pioneers would blaze the way
for how observing's done today.

Frederick Scott Archer's wet
collodion process (*c.* 1813–57)

A photograph (or CCD)
records more than the eye can see,
and with immaculate precision.
If they had worked with human vision
instead of silver's instant tan,
they'd still be back where they began,
peering through prisms at a smear
whose message was by no means clear.
But modern ways of trapping light
mean stargazers can sleep at night.

9 On to the Beginning
(after 1900)

The Ladies of Harvard

Among the mass of names to choose,
few *ladies* have inspired my Muse,
but here are two to dwell upon –
Annie, and Henrietta Swan!
At Harvard College, USA
(a major centre of the day
for practical investigation)
they both acquired a reputation
that's now in undeserved decline,
and so I hope these words of mine
will help to reinstate their names.
First, let's attend to Annie's claims . . .

Annie Jump Cannon 1863–1941
Henrietta Swan Leavitt 1868–1921

1 MISS CANNON'S NEW WORD

Checking the spelling . . .

She was a Science graduate
(her deafness helped her concentrate),
when Physics wasn't quite the thing
for Ladies. Edward Pickering,   1846–1919
Harvard Observatory's Director,
was motivated to select her
thanks to some crazy scheme of his
called 'Equal Opportunities'.   1896
There was a project he was starting
for photographically charting
star spectra; so she worked at night
(the Moon had to be out of sight)
with negatives like window glass.
Each star's specific spectral class,
revealed in its little smudge,
she had to analyse and judge.
She chucked her boss's old ideas
(he'd sorted all the different smears
into an order ABC . . .),
proceeding, rather cheekily,
to go by *temperature* instead –
from hottest (white) to coolest (red).
The sequence, by this stratagem,
was OBAFGKM

(some classes lost, some rearranged).
This series has remained unchanged,
inviolate as Holy Writ,
though others have extended it.   Now WOBAFGKMRNS
Two astrophysicists took note
of this strange word the spectra wrote . . .

The Hertzsprung-Russell Diagram

Both Ejnar and Henry decided to see*
how Miss Cannon's new classification
(which measured how *hot* every star had to be)
went with *L* (that's their *light radiation*).†

The O and B types, which are hottest of all,
they found are exceedingly bright.
A, F, G, K and M all progressively fall
on a line that sinks down to the right.

This is called the *Main Sequence*. The Sun's in this lot –
(it's a star of type G, by the way).
Most stars belong here, but some odd ones do not –
so what law do these rebels obey?

*Ejnar Hertzsprung 1873–1967, Henry Norris Russell 1877–1957
†Luminosity

Take those M stars with *L* like the O's and the B's!
Since their surface emission is low,
they've got to be huge to compete with all these –
they're *red giants*, a name you will know.

And what about these, in Miss Cannon's Class A –
stellar glow-worms compared with the rest!
They've got to be tiny to shine in this way . . .
they're *white dwarfs*, as you've probably guessed.

The Diagram, everyone knew, held the key
to the secret of stars' evolution . . .
But decades went by before A, B and G   See p. 207
wrote their paper that gave the solution!

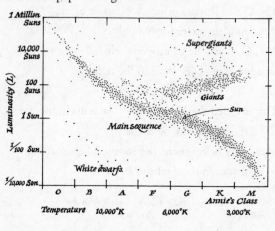

2 MISS LEAVITT
PLUMBS THE DEPTHS

Head in the Clouds

While this research was going on,
Miss Leavitt (Henrietta Swan),
like her friend Annie hard of hearing,
was also earning money peering
at photographic negatives.
Our galaxy's small relatives,
the Magellanic Clouds, both lie
towards the southern polar sky –
the goodies of which they're possessed
make them a stellar treasure chest.
Miss Leavitt checked each stellar dot,
discovering that quite a lot
were *variable* (not so bright,
or brighter, on a different night).
The fact that they were flickering
intrigued Professor Pickering,
who told her she should concentrate
on seeing how they oscillate.
She measured hundredweights of glass,
and came up with the famous class
known as the *Cepheids* . . . the key
to distance in Cosmology!

Published 1912

[193]

Henrietta's 'tape measure'

Dear Friend,

        Each Cloud is quite compact,
from which derives the useful fact
        that every single star
is more or less as far away
as all the rest — so I could say
        that *how they seem, they are!*

Let me explain . . . *A* outshines *B*,
by some known factor, let's say 3,
        when I have measured it.
But since they're side by side, we know
that this must be the ratio
        of *L* — what they *emit!*

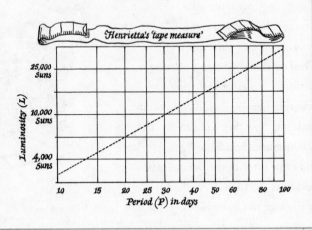

I found a group of stars, whose light
was altering from night to night
      in ways that were unique.
The bright ones might take fifty days
to go through their entire phase —
      the faint, less than a week!

As my name's Henrietta Leavitt,
I swear, I hardly dared believe it,
      it was so unexpected!
Their L (or luminosity)
and P (their periodicity)
      are definitely connected!

You get the point? Elusive π
can be derived for stars nearby —
      for far ones it's too small.
To measure distances much greater
we need a new discriminator
      with no constraints at all . . .

Now, if we know a star's true L,
it isn't difficult to tell
      how far off it must be
(L says how bright the star appears
when seen from 33 light-years).
      With my stars, measure P . . .

... and *L* is easily worked out
That's what my method's all about –
    nobody's found one better!
You'll learn on a succeeding page
that *Edwin* used it as a gauge!

Sincerely, Henrietta.

PS: Some Cepheids were spied
quite near the Sun, and duly π*ied* –
    so *L* was found precisely.
These values firmed up all the rest,
and then my brilliant distance test
    was calibrated nicely!

The Master Builder

From this time on, the USA
was where most of the action lay.
It's due in part to clearer skies,
but mainly to the enterprise
of George E. Hale, physicist,   1868–1938
whose patter no one could resist.
How many purses, swollen tight,
unzipped to help his cause (*More light*)!
He built a 40-inch refractor –   1897
Charles Yerkes was its benefactor,   1837–1905

one of Chicago's millionaires,
who made his pile from streetcar fares
and cheerfully met every bill,
though George was in his twenties still.
More money from the private sector
endowed a 60-inch reflector,
so he set off at once, to seek
a good place on a mountain peak.
He trekked up trackless slopes; of these,
Mount Wilson, near Los Angeles,
turned out to be the best for viewing.
(His colleagues asked what he was doing,
selecting such a wild spot.
A monk might like it – they would not!)

Hooking Hooker

The 60-inch was in commission,
but lost its premier position
when John D. Hooker* had a visit . . .
The skies, George told him, were exquisite,
and so the fact had to be faced
that 60 inches was a waste –
100 inches sounded right
for such an admirable site!
So Hooker drew out of his bank
sufficient for a mirror blank,

*c. 1849–1911

Carnegie making up the rest.   Carnegie Foundation
It's 1918 . . . You'll have guessed
that Hubble's waiting to appear.
But first, let's get a few things clear!

The Great Debate

When George trudged up his mountain track,
the ghost of Herschel at his back,
some questions that his mentor put
while standing at the 40-foot
remained unanswered. Might there be
more space beyond our galaxy
containing other galaxies,
or is ours Everything That Is?
These questions fuelled a Debate
that's still referred to as 'The Great' . . .

———

Venue: National Academy of Sciences

Date: 26 April 1920

Advocates: HARLOW SHAPLEY (1885–1972)

HEBER D. CURTIS (1872–1942)

THE GREAT DEBATE

SHAPLEY

I'll begin the debate, if I may?
Some of Miss Leavitt's stars, far away,

give the *Via Lactea*
(let us use the light-year)
a width of some 300k.   Three times too large.

CURTIS

That's grossly too large! In addition,
did you say in your formal submission
that we are unique,
and it's pointless to seek
other galaxies?

SHAPLEY

                    That's my position.

CURTIS

I suggest that those nebulous spots
(of which there appear to be lots)

could be galaxies too,
especially a few
that are spiral, with luminous knots.

SHAPLEY

But van Maanen has claimed they rotate!   1884–1946
[Which all subsequent work would negate.]
How could they be vast
if they're turning so fast?
No – they're stars in an embryo state . . .

. . . in a gas cloud that's steadily spinning,
like the Sun when it had its beginning.
So they're not at all far
from our neighbourhood star.
(Do you know, I've a hunch that I'm
    winning!)

CURTIS

In Andromeda, near the star μ,
there's a naked-eye nebula.   The Andromeda Galaxy

SHAPLEY

                    True.

CURTIS

From its size, I would say
about 500k
is its distance from me (and from you).

SHAPLEY

You mean it's a galaxy? *What?*
It's a localized luminous spot!
If that's how far it lies,
it's a *tenth* of the size
that I gave for our galaxy! Rot!

CURTIS

I have to admit I'm impressed
how you worked that out! Yes, I suggest
that we're 30k wide.   Three times too small
Time, of course, will decide,
but I'm certain I've come out the best!

SHAPLEY

Dr Curtis, I can't accept that!
You have talked, my dear sir,
    through your hat.
As you'll recognize later,
when faced with more data . . .

CURTIS

Well anyway, thanks for the chat!

Speed check

An expert in spectroscopy
had looked at certain nebulae
ten years, or maybe more, before
our two contestants took the floor.
This pioneer was Vesto Slipher,   1875–1969
who wondered what he might decipher
about their temperature, etc.
The lines he measured in their spectra
were shifted from their true location,

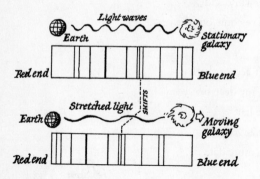

which was a well-known indication
of rapid motion in the source
(that's relative to us, of course).
The nebulae that Slipher checked
were similar in one respect —
their distances were all *increasing*.   William Campbell (1862–1938)
Others took part in this policing,   James Keeler (1857–1900)

clocking their speed, but not yet knowing
where these mysterious things were going!

Edwin climbs the mountain . . .

In 1919, George signed up
the winner of a sporting cup
who'd taken a degree in law
and rose to Major in the War.
Meet Edwin Hubble, Ph.D.!   1889–1953
He shortlisted some nebulae
to see if he could find how far
these tantalizing objects are.
Working with Milton Humason   1891–1972
(a caretaker whom George took on,
and who became distinguished too),
Miss Leavitt's stars came into view,
their friendly rays now bright, now dim,
like lighthouses informing him
that he would shortly come to port
in realms beyond the reach of thought!
From careful measures, he could tell
their P, and thus work out their L . . .
900,000 years of flight   Because of details he lacked,
had marked the passage of the light   it's more than twice as far, in fact.
from Heber's prototypal blur
that shimmered in Andromeda!

Shifting . . .

Do I risk ridicule to say
that Ed and Milton swept away
the last sense of self-centredness?
Our status had grown less and less . . .

The Earth had lost its pride of place;
the Sun was no great shakes in space;
and now the galaxy was just
another whirl of cosmic dust –
not All That Is, but One In Many.
In other words, we're two a penny!

Carl Wirtz had meanwhile suggested,
based on some nebulae he'd tested,
that red shift (or velocity)
is linked to their profundity –
the fastest are the most remote.
The landmark paper that he wrote   1923
could mean that space is getting bigger!
This was a most effective trigger
which launched our astro-
      mountaineers
towards the peak of their careers!
To cut a tangled story short,
they came up with a term (H_O),
the *Hubble Constant* (still disputed),
by which astronomers computed
an object's *distance* from its *flight*,
using the graph that's on the right.

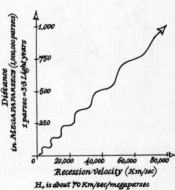

H_o is about 70 Km/sec/megaparsec

Einstein writes a letter . . .

When Albert Einstein had the nerve   1879–1955
to say that Space and Time both curve   General Relativity 1915
(five years before the Great Debated),
Willem de Sitter calculated   1872–1934
that space is probably inflating.
To Albert this was irritating.
He thought the Universe unchanging,

[205]

and made it stable by arranging
to have a Constant introduced,
denoted λ,* which produced
the proper balancing effect
(or ought to, if it was correct).
This suited his beliefs much better.
However, this contentious letter
did not, in fact, last very long,
since Ed and Milton proved him wrong –
their findings on the mountain top
sent λ flying, neck and crop!
But Albert never really cared
for Big Bangs – feelings Hoyle shared.

*The Cosmological Constant

Newton had thought the
balance odd,
but his own λ term was God.

Fred Hoyle 1915–2001

αβγ soup . . .

I've jumped some twenty years ahead
by introducing 'Steady' Fred;*
but facts accrued at such a pace,
I wish this book could mimic space
and stretch each page out as I go
(making the clock tick nice and slow).
By no means everyone agreed
that Slipher's shifts denoted speed,
or that the Universe could start
because a pinpoint blew apart,
and not till 1948
did that renowned triumvirate
of Alpher, Bethe and Gamov
succeed at last in pulling off
a triumph for the Big Bang cause.
Applying well-established laws,
they worked out how to follow through
the cooling of the Cosmic brew . . .

*He thought that matter was created
to fill up space as it inflated.
He called his scheme the Steady
State . . .
support for it was never great.

Ralph Alpher b. 1921
Hans Bethe 1906–2005
George Gamov 1904–68

Atoms developed as it chilled,
until vast tracts of space were filled
with *hydrogen* (90%)
and that unearthly element
called *helium*. Both these abound
in just the ratio they found!

This took about 3 minutes

Recycling

But these two are no recipe
for planets, plants, or you and me!
More elements were needed later
to meet the needs of the Creator,

[208]

and this was Hoyle's inspiration —
they were produced by transmutation
within a turbulent *supernova*,
a star whose life is almost over.
The heat inside is so intense
it generates new elements,
and having done so, blows apart!
So life (and planets) couldn't start
till supernovae spread enough
of this regurgitated stuff
to make the Earth on which we stand.
We're definitely second-hand . . .

Stellar nucleosynthesis 1946

How the Sun shines . . .

It's hard to rhyme this, but I'll try . . .
Within a star, two nuclei
of basic hydrogen become
one nucleus of helium,
whose mass is very slightly less.
It isn't difficult to guess
what this means, since you've been prepared
for E to equal mc^2!

Four million tonnes of Sun per second
(so astrophysicists have reckoned)
are being lost, and turned to E –
this can't go on indefinitely . . .

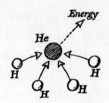

Highs and lows

Remember Messrs H. & R.
(you needn't look back very far)
who drew a Diagram or Plot,
with every star shown as a dot?
Most of the stars, although not all,
lie in a Sequence, you'll recall.
It seemed a reasonable guess
that stars would gradually progress
along this line as time went on –
but this was scotched by Eddington!
Their *mass* decided where they lay –
and at that point a star would stay
until it claimed its pension rights!
So O and B, the leading lights,
have much more mass than dimmer G
(the class our Sun is known to be).
Soon, astrophysicists could trace
how aging in a star takes place . . .
They hold their own till they begin
to sense that fuel stocks are thin;

See p. 191

Sir Arthur Eddington 1882–1944

Five thousand million years,
    maybe,
before the Sun's in jeopardy.

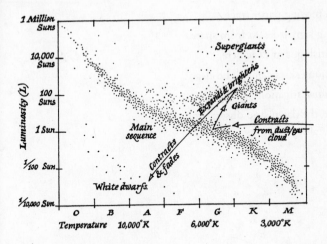

and when this happens, as a rule
they puff up and grow huge and cool
(*red giants*, to the upper right),
maybe a hundred times as bright.
They then collapse, and end their days
tiny and hot (the *white dwarf* phase),
when they're fantastically dense.
I hope this brief account makes sense . . .

Listening in

A by-product of World War 2
was *radar*; and to this is due
Cosmology's next giant stride,
when pens were jerked from side to side

across a moving paper trace,
recording signals heard from space.
The British got in early here,
since cloudy skies can still be clear
to radio investigation –
nor does the dawn stop observation.

Sir Martin Ryle's* catalogues,
drawn up in reclaimed fenland bogs
near Cambridge, listed the position
of sites of radio emission –
one source (3C 273)
was narrowed down, eventually,
to what looked like a distant star.
The giant eye on Palomar*
sifted its light . . . Its red shift proves,
given the speed at which it moves,

*1918–84

*The 200-inch (1948)

that some *two billion years* have passed
before its beams reach us at last!
The archetypal QSO*   *Quasi-Stellar Object
had been unearthed by radio.

Enter black holes . . .

At first, their distance was in doubt,
since physicists could not work out
how they could blaze away so brightly.
Some sceptics had observed, politely,
that *gravity*, when it's intense,
produces spectral measurements
that look as though they're caused by speed —
if this was so, there was no need
to stretch the rule of Hubble's Law
to justify the shifts they saw!
But nowadays it's pretty clear
they're bright and far, not faint and near!
Around a black hole (which is thought
to be the supermassive sort),*   *A mass up to a billion suns
a swirling cloud of gas rotates   is likely for the biggest ones.
and gradually accelerates   The Milky Way has got one too:
until it turns into a fuzz,   it's peaceful now, and out of view.
and radiates (as quick stuff does)
prodigious quantities of light,
explaining why they seem so bright . . .

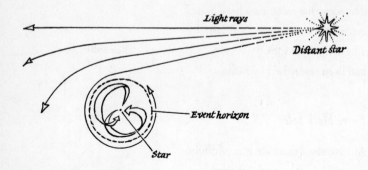

Enwrapped, entrapped . . .

Black holes (the topic of the hour)
can be a useful source of power
because they are extremely dense,
and so their gravity's intense.
The densest medium around
is *neutron* essence, which is found
inside an atom's nucleus . . .
Compress each atom that's in us
until the waste of space within
is flattened like an empty tin,
and not even our friends could spot
the sweet sub-microscopic dot
that is our insubstantial All!
You'll read below what may befall
a star whose fuel stocks grow thin:
eventually it just caves in,

its neutrons fuse, and we set eyes on
a thing called an *event horizon*.
Its *g*-forces have now enwrapped
its space around it, and entrapped
its light within — so all we see
is *nothing!* But its gravity
can still attract, and it may grow
into a proper QSO
if it expands itself enough
by sucking in surrounding stuff.

THE STAR THAT GOT TOO HOT

Main Sequence stars like diamonds lie
upon the velvet of the sky;
but She who is about to die
illumines, from her place on high,
    the Hertzsprung-Russell Plot.

She is a Supergiantess;
the Lady's mass (if one may guess)
is fifty of our Suns, no less;
      but cheerful she is not.

'Alas!' she cries, 'my precious store
of hydrogen will last no more!
I feel nervous at my core;
it's grown much larger than before,
      and it is far too hot.
My helium will now commence
creating other elements;
the situation's very tense –
      how little time I've got!

'Let me explain the situation.
I'll be puffed up by radiation
because of all this transmutation;
but when I start on the creation
      of *iron*, that's my lot!
For iron saps my energy,
and my colossal gravity
will get a real crush on me
      and squash me to a dot.

'Then will your eyes behold a sight!
My immolation will ignite
a blaze that seems at least as bright
as all the stars you see at night
        (which come to quite a lot).
But very soon, it will subside;
and as I shrink from side to side,
my curse will come upon me!' cried
        the Lady. 'Doubt it not!

'As all my quarks coagulate,
my density will get so great
that I'll become *degenerate*.
In this unprecedented state

the gravity I've got
will mean I'm in it up to here.
My space will curve into a sphere,
and I'll completely disappear –
which puts me in a spot!'

The Lady's fears were all too true.
A supernova came in view,
and people stared, as people do,
at something that is briefly new;
but she was soon forgot!
A black hole opened at her feet;
her odyssey was now complete;
but somewhere, does the heart still beat
of She who got too hot?

Beyond the limit . . .

You will expect me to be talking
about Professor Stephen Hawking –   b. 1942
the man whom everybody mentions
when contemplating new dimensions!
Alas, my Earth-bound 3-D brain
has snapped its synapses in vain,
trying to get into my head
a fraction of the things he's said,

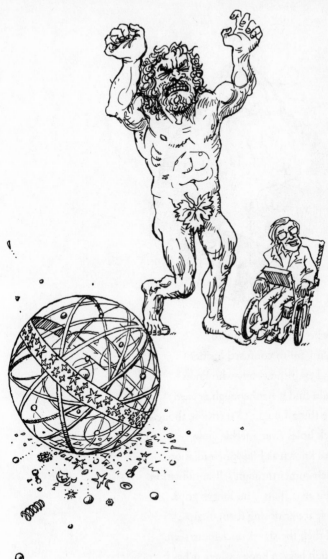

so who am I to be a guide
when I am so confused inside?
(And even somebody who knows,
would find it hard enough *in prose!*)
One thing I ought to mention, though . . .
Black holes emit a feeble glow
that's known as Hawking radiation,
which comes through self-annihilation.
Their mass falls, *g* no longer grips,
the space enclosing them unzips,
and then the star is once more seen,
though heaven knows where it has been!

The infinite jigsaw?

Where is the Cosmic story going?
The aim to be all-wise, all-knowing,
is laudable — but bear in mind,
that there is always more to find!
Less than a century ago
astronomers still didn't know
how stars shine; and not knowing this
caused general paralysis —
the task was foremost on the list
of every astrophysicist!
That problem's sorted out all right;
but now another one's in sight,
for some old stars should have been born
before that scintillating dawn

when Time began; and, without clocks,   14 billion years ago
that's something of a paradox.
It will be sorted out, of course,
perhaps by some expanding force
that puts Creation back a bit;
but making all these details fit
could bring new problems into view . . .
If each one answered raises two,
then what appears to be success
may really mean that we regress –
the more we know, the less we know!
If this is the scenario,
the knowledge jigsaw will increase
each time we add another piece;
so, till the last one is in place,
take my advice, and

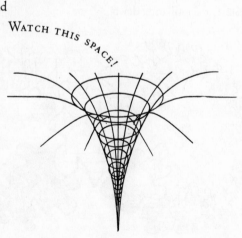

WATCH THIS SPACE!

Epilogue

The Man in the Chair

Says the Man in the Chair
(whose thoughts we can share
through ingenious vocalization),
the Universe might
have gone out like a light
a split-second after creation!

That it didn't, is clear
from the fact that we're here;
but it doesn't mean God was about,
for sooner or later
(without a Creator)
a Big Bang was bound to work out!

So no effort was needed,
and the Cosmos proceeded;
but I'd welcome one piece of advice.
If Chance was the key
to the wonders we see,
please tell me — whose Hand rolled the dice?

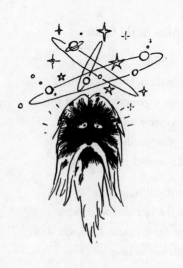